ENDORSEMENTS

Wow, the moment has come. *The Fight to Lead,* and you have done exactly that. Armon Owens, Master Chief, my brother, and my friend have made me extremely proud. Armon has excelled in his ability to captivate an audience, compel them to pay attention, and learn from his life experiences. Although many people open their mouths and let words flow, it is unusual to encounter someone who will actually allow those words to come from their heart to make you a better, more complete person. Armon is a transparent man. He shares his joy and suffering, highs and lows, triumphs and failures. Enjoy the page-turning experience and give yourself the freedom to develop.

— Marte Gifford
Retired Staff Sergeant Army

In my opinion, *authenticity* is a key, if not *the* key, in leadership. Armon Owens is as authentic as a leader can get. His care for those he is charged with leading is genuine. The person shows through daily as he walks his walk and talks his talk. When there is genuine trust in those who lead, there is substantial buy-in from those being led. I can say beyond the shadow of a doubt that Armon is as real as they come, and I implicitly trust his leadership and judgment. Armon is my friend and my brother, and I am more than happy to endorse anything he has his hands on!!!

— Adrian Watkins
Command Master Chief, Submarine Qualified

THE FIGHT TO LEAD

How To Overcome The Battle Within To Lead At The Highest Levels

By

ARMON OWENS

Paperback: 979-8-9879184-1-8

Hardback: 979-8-9879184-3-2

E-book: 979-8-9879184-0-1

(LCCN): 2023903777

DEDICATION

To my life—you are my breath, my everything.

Ty, thank you for believing in me when I didn't believe in myself.

To Armiya, Justice, and Faith, you are my strength and my WHY. Thank you for being you.

To my first fan, my Mom, thank you for your supernatural encouragement in pushing past wherever I could imagine. Thank you.

To Amanda Morris May, you were the first person to tell me that I was meant to write. I am still living off of your encouragement thirty-five years later.

Thank you to anyone that has read anything that I have written.

I appreciate and honor your time.

TABLE OF CONTENTS

FIGHT TO LEAD - FOREWORD

"If the only tool you have is a hammer, you tend to see every problem as a nail."

~ ABRAHAM MASLOW

L eadership cannot be taught.

Let me clarify. Leadership SKILLS can be taught. Things like managing expectations, fighting your ego, and managing your emotions in stressful situations. These things can be taught, and they are worth significant study by anyone desiring to be a leader of consequence or someone finding themselves suddenly thrust into the spotlight of leading.

What cannot be taught is the ability to withstand the heat that comes from being in that spotlight. You have to want to be the person in charge or at least be willing to shoulder the burden that comes from being given that level of responsibility.

If you are that person, this book is for you.

Being willing to carry the load of leadership is one thing; being able to do it with style and grace is another. To do that, you need a variety of tools in your leadership toolbox. Many of these tools aren't about dealing with the people you lead—they are necessary to deal with yourself.

You must be able to lead yourself before you can effectively be a leader of others. It is a common phrase in the leadership space—in order to lead, you must be worthy of being followed. Creating that worth starts with self-reflection, growth, and winning the battle within yourself.

Make no mistake; you need to have all your stuff together to be a leader because the heat from that spotlight will not only show all your cracks and flaws like the harsh light of florescent lights in your bathroom, it will work its way into those cracks and make them bigger and bigger until you fall apart.

But here's the thing, we all have those cracks. The goal is not to get rid of all your flaws; it's to create awareness in yourself that your flaws exist and constantly work on them. Working to improve how your flaws affect your performance, continually working to ensure they don't accidentally de-rail you, is a lifelong pursuit.

We are who we are. The question is, what are you going to do with what you've been given? This book is here to help you with that. It's here to give you various tools to deal with your various flaws. After all, different problems have different solutions, and you have to apply the correct tool to get the result you desire.

* * * * * * * *

I had the once-in-a-lifetime honor of being in command of a nuclear submarine in the United States Navy. Make no mistake, a submarine at sea, submerged, is a leadership laboratory as unique, demanding, and pressurized as any you will find.

My job was made immensely easier by getting to work with a once-in-a-generation leader like Armon Owens. Armon was my senior enlisted advisor, the Chief of the Boat—affectionately referred to in the submarine force as the "COB."

I worked with Armon for almost three years as we led the USS HELENA on a remarkable run of success. Leading a nuclear submarine crew is an awesome honor and tremendous responsibility. Not just responsibility for the ship itself—we'd been given a two billion dollar asset from the taxpayers after all.

More importantly, we'd been given 160 sailors sons, brothers, and fathers all, to lead, mentor, and help grow. At the end of the day, a submarine is just a hunk of steel with wires, electricity, and pressurized fluids running through it. It is the people who make it run. It is the crew who brought success to the name HELENA.

Armon and I were keenly aware of our responsibility to ensure the ship was successful in its mission—which is to always be ready to provide credible combat power to the greater Navy. But we also wanted to ensure that we were successful in our personal mission, which was to make each one of the sailors onboard our ship the very best person they could be.

That started with us understanding ourselves and then understanding each other, so we could work together effectively to take the rest of the crew to levels even they did not think they could achieve.

To say we took them, there is a bit of a misconception. Was the ship extremely successful? Yes. Did the crew achieve immense success? Absolutely. Every senior enlisted leader on that crew was the top performer in his position out of a squadron of 15 other submarine crews.

Every department head onboard moved up to Executive Officer. Every Executive Officer, the second in command, has gone on to

Command their own submarines. The ship earned the top award for battle efficiency in the squadron—twice. And we earned a meritorious unit commendation for our deployment in 2017.

I attribute most of the ship's success to Armon. If I have one skill as a leader, it was recognizing that I had a racehorse in the form of Armon, so my job was to get out of the way and let him run. He is an extraordinary leader because he is thoughtful, caring, and tenacious. He expects the best of everyone around him, but he is also dedicated to helping you get there.

It's a rare thing when you, as a leader, get someone on your team that makes you stop and say, "Oh, ok. I'm going to have to step up my game here." Armon is that kind of transformational leader, and I knew it was on me to make sure I gave him my very best. To do any less would not only waste his talent, but it would also rob my crew of the chance to learn from him, and that would be unconscionable.

So, did Armon and I take the crew to their highest level of performance? Absolutely not. As leaders, we are merely guideposts on the road to success. You cannot make people succeed—you can only establish the conditions and then show them the way.

They have to walk it themselves. But they won't even consider walking the path unless they believe in where you are trying to guide them. Again, you have to be a leader worthy of being followed. We didn't take them there; they got there on their own—but Armon gave them the tools they needed to do the work. To work on themselves so they could be better leaders, better Sailors, and better people.

Armon has bottled his knowledge and philosophy into this book. To show you, through his experiences and those of others, how to do this work on yourself.

> Can you put your ego aside and focus on your team?
> Can you acknowledge your fears and put them aside to be the strength your team needs?
> Can you do the work on YOU, so you can lead in a way that is not about YOU?
> Can you win the battle within yourself?

It's a lifelong journey on the path Armon is about to lay out, but it starts with the first step.

You owe it to yourself to do this work. Those you lead deserve the best YOU that you can be. Armon will show you where you need to go, trust me.

Let's get to walking...

—Captain Jason Pittman
Commodore, Submarine Squadron 6

INTRODUCTION

A grandfather was sitting with his young grandson to give him some wisdom. The grandfather explained to his grandson that there are two wolves in constant battle within every person. The Bad Wolf is full of jealousy, anger, regret, and fear. Meanwhile, the Good Wolf is full of hope, happiness, love, and faith. The young boy asks his grandpa, "who wins?" To which the grandfather replied, "the one you feed."

In 2015, I had the honor of reporting as Chief of the Boat to the finest submarine in the world at the time, the USS Helena (SSN-725). After seventeen years, I had the privilege of leading my own team of submariners into harm's way. I dreamt about this moment from my very first submarine, USS Oklahoma City, in 1999.

I was inspired by the presence and leadership of my Chief of the Boat, affectionately known as the *COB*. I knew that one day I would be a *COB*, and today was that day. I was on cloud nineteen, getting dressed that morning, preparing for my first time at sea as the COB, and excited for the opportunity to serve.

When you finish the demanding requirements associated with becoming a COB, you are allowed to wear a badge on your uniform indicating that you are the COB, called a *cookie*. Looking at myself before leaving for work, I thought that cookie looked nice under the light. Not only would I do a great job, but I would also look great doing it.

After a few days at sea, the time comes for one of the COB's favorite events, *field day*. I know you're thinking, *how can you possibly do a field day on a tiny submarine at sea?* Well, the Navy has a twisted definition of a *field day*. *Field day* in the Navy is when the entire crew comes together to clean the submarine. The absolute opposite of the *field day* you were thinking of, right?

This was a big deal for me; I had waited for seventeen years to lead my very own *field day* as the COB. As the COB, we are responsible for the cleanliness and preservation of the submarine. When you purposely submerge a big black tube with a nuclear reactor, everything must remain pristine for operations to go successfully.

Usually, the COB walks around to ensure everyone is cleaning effectively and up to standard, but I wanted to clean with my crew—to get down in the bilges with my team and fight with them.

After we finished, I felt like field day was an absolute success until one of my Sailors pointed out the *U* was missing from USN on my cookie. I looked down, and immediately, I was furious. I didn't have another cookie to replace the broken one. I was beside myself because how would I stand before the people I led with a noticeable fault?

Frustrated, I went to sit in my office and threw the cookie on my desk. Held captive by my anger starring at what I considered an abomination, I remembered a quote from Jim Carrey, "In life, nothing happens to you; it happens for you." Jim was conscious of perceiving challenges as beneficial so that he could deal with them most beneficially.

From the perspective of this incident happening *to* me, I saw this moment as a message, not a detriment. Every day in our lives, we have to fight the innate pull to be self-serving and see the world revolving around us.

In leadership, this fight doesn't cease, it only gets more challenging, and the pull becomes stronger. You are the one in charge, you have the power, and you are the one responsible for accomplishing the mission. If it fails, it's on you. You, you, you.

When you lead a team, *you* are merely part of the whole, and for success to happen, the whole has to perform, not just *you*. Leadership is about the value you can provide for others. The leader resembles the composer; they must turn their backs on the crowd, the accolades, the adulation, and the applause. They must turn their backs on the things that feed their ego and turn toward the things that feed the people they serve.

Do you have the capacity to facilitate an environment where everyone on your team can be the best version of themselves? Do you have the power to confront any limitations within you that may hinder others' ability to meet or even exceed their potential?

Over my twenty-six years of naval service, I experienced self-centered interactions fueled by a need to dominate and control. Early on, I assumed that this was what leadership was. However, leadership does not equal control.

The best leaders know that their role is not to *dictate* but to *inspire* and motivate others to act. That means they know how to conquer their personal battles and step out to lead. These are the two right hooks: Conquering internal battles and not dictating by inspiring others. Leaders, surrendering the idea of control invites your team to discover their potential truly, thus increasing the probability of success.

Now, remember, I was so concerned with myself that I completely missed the point. I was so troubled by my appearance that I lost sight of my true goal. The Challenger won that round, but I wiped the sweat off my face, patched up my cuts, took a swig of water, and jumped back in the ring!

The *U* being stripped off of my cookie wasn't a mistake. It was an invitation to an opportunity. An invitation to explore how significant others could be if I focused on adding value to them rather than feeding my ego. The opportunity to remove the *U* from my leadership equation means that the people I have the privilege of leading may be made whole.

To release our full leadership potential, we must remove the *U*. We must be willing to confront the limitations within us that will most assuredly limit the possibility of those we lead. This undertaking is the fight of a lifetime because it genuinely lasts a lifetime.

We naturally gravitate toward the things in our lives that help us avoid pain or please us. The journey of leadership consists in denying *self* and elevating the concern for the conditions of *others* above *yourself*. This fight for leadership involves running toward the pain and sacrificing immediate pleasure.

The Bad Wolf within us has the loudest roar and is addicted to fulfilling only its own needs and wants. The Bad Wolf is self-serving and only looking to consume for consumption. Left to its own devices, it will ultimately consume itself. The fight to lead is the fight against that Bad Wolf. To starve the Bad Wolf, so its roar becomes a whisper and your true strength to lead is released.

This is the fight that some of history's most outstanding leaders have participated in, from Nelson Mandela enduring the pain of twenty-seven years of imprisonment, overturning apartheid, and becoming the nation's first black president to Abraham Lincoln nearly dying during a duel to leading the Union through the Civil War.

The Fight to Lead outlines the necessity to remove the "*U*" through stories from my own life and the lives of some of the most significant historical figures.

The Fight To Lead is not an attempt to be the *definitive* work on leadership, and *The Fight To Lead* does highlight the *definitive work on oneself* that is required to be the leader that people genuinely need.

ROUND 1

THE CHALLENGER—BELIEF

The only person you are destined to become is the person you decide to be.

—RALPH WALDO EMERSON

One young man has his bike stolen and seeks revenge…

One young man is cut from his high school basketball team and is devastated…

Publishers tell one young author fifty-one times that his book doesn't cut it…

A doctor is thrown into the concentration camps by Germany…

A hockey team is embarrassed by their rival right before the Olympics…

Her life's work was rejected sixty times…

The above stories are not about the beginning but about the end. In the end, the kid who had his bike stolen grew to be the greatest boxer of all time—Muhammad Ali. The young man cut from his high school basketball team grew to become the greatest basketball player—Michael Jordan. The author who was rejected fifty-one times is the current New York Times Bestselling author—Andy Andrews.

The doctor subjected to one of the most horrific events in human history was the author of *Man's Search for Meaning*, Viktor Frankel. The team that was embarrassed by the Soviet hockey team in 1980, right before the Olympics, was the gold medal-winning USA hockey team. The young author whose life work was rejected sixty times before someone published her was Kathryn Stockett, New York Times bestselling author of *The Help*.

Many of us would understand if they quit under the circumstances they faced—who wouldn't? We look at circumstances as the arena of our lives, and leaders rise to meet them and take them head-on.

Being a leader is one of the most challenging and overwhelming positions to serve in. It isn't something that comes naturally; it is something that you have to fight for. The first fight you must overcome to become the leader your people need you to be is the fight to believe.

Believing in yourself, your team, and the purpose of the organization you serve is always first. Denis Waitley said, "If you believe you can, you probably can. If you believe you won't, you most assuredly won't. Belief is the ignition switch that gets you off the launching pad."

There is a belief in all great leaders that drives them never to give up, and the belief is in something beyond themselves. Regardless of your religious background or faith, there is a scripture in the Bible that is very fitting for the leader, "We walk by faith, not by sight."

Leaders don't know the ending, and there is nothing easy about believing in something that you can't see; however, just because it's not easy doesn't mean you can't do it.

We've all heard of the story of David and Goliath. Despite the extraordinary odds that David was up against his faith, his belief became more significant than the obstacle or giant in front of him. With the conviction that grew from that belief, he took action, and the outcome was incredible.

What makes someone keep going after sixty rejections for the book *The Help?* Most people can fight for three, maybe a dozen, but what makes your belief strong enough to push you to keep going after sixty rejections? Is sixty-one a magical number? The number was irrelevant, and it was the belief that no matter what, the author's conviction of the story was more substantial than the rejections.

As a leader, fulfilling the purpose of the mission or the operation, the project must be more substantial than the fear of failure. That commitment is leadership. These phenomenal leaders wouldn't allow the beginning of their stories to become the end of their stories. They kept going, and they kept persevering.

Many things have been written about the phenomenal individuals listed above, and entire books were written about their work ethic alone. When asked how they overcame the circumstances to achieve such greatness in countless interviews, the response would primarily be, "I just kept working, kept trying, and then one day...it happened. It's all about hard work!"

Speeches, books, and TED Talks from the great ones proclaim that the secret to greatness is to work hard. On the surface, it would appear to be true. Over and over again, we are told something similar: work hard and that everything we desire will be yours. If you aren't where you want to be, you *haven't worked* or *aren't working* hard enough. Hard work is merely what you see, and it's the easy answer.

A few levels deeper, below the surface, out of the limelight, and invisible to the masses, lies the true catalyst of their success. Amidst all the evidence and the voices speaking contrary to their success is a whisper emanating from the depth of their soul. A whisper refusing to be drowned out by the contrarians and underachievers, and a whisper that refuses to be silenced by the overwhelming odds. A whisper that won't go gently into that good night. A whisper that confirms who you are when you succeed.

Some call this whisper your consciousness or your inner voice. This voice is the resonance of our beliefs. Whether you win or lose, the voice of belief within you is the determining factor. Our ship's rudder is our inner gimbal synchro directing us towards success or failure.

Whether an audible voice or an intuitive pull, it's that whisper that knows the truth. The truth is that this difficulty will be the most significant thing for you once you get to the other side. The truth is that the most significant leaders have all faced the greatest challenges. To be a true leader, you must not only endure pain but choose to subject yourself to that pain willingly.

Look at what Muhammid Ali did. Born Cassius Marcellus Clay Jr., he grew up in the south in the 1950s and learned to commit to his convictions and stand his ground. In fact, at the age of twelve, Clay discovered his talent for boxing through an odd twist of fate. The event leading up to this great discovery was his bike being stolen.

Clay spoke with the police about his bike, and in his frustration, he told the police officer, Joe Martin, that he wanted to beat up the thief. Clay was committed to finding the thief and teaching him a lesson, but Joe Martin saw the flaw in this plan and said, "Well, you better learn how to fight before you start challenging people."

In his autobiography, Ali said, "I ran downstairs, crying, but the sights and sounds and the smell of the boxing gym excited me so much that I almost forgot about the bike," The thin twelve-year-old was captivated by what he saw.

"There were about ten boxers in the gym, some hitting the speed bag, some in the ring, sparring, some jumping rope. I stood there, smelling the sweat and rubbing alcohol, and I felt in awe. One slim boy shadowboxing in the ring threw punches almost too fast for my eyes to follow."

In that gym, Clay completed the stolen bike report and was about to go when Joe Martin decided to change the course of Ali's life. Martin tapped him on the shoulder and said, "By the way, we got boxing every night, Monday through Friday, from 6 to 8. Here's an application in case you want to join the gym."

Fast forward a decade and a half later, on Feb 25, 1964, Muhammad Ali faced the fight of his life. Relatively unknown to the greater boxing world, he faced 7-1 odds against the world champion, Sonny Liston.

Muhammad Ali is considered by most to be the greatest heavyweight boxer of all time. He is a three-time World Champion responsible for some of the most epic fights in boxing history, from the "Rumble In The Jungle" to the "Thrilla In Manilla," he thrived in the spotlight when many boxers allowed their managers to speak for them.

Ali was named the Sports Personality of the Century by Sports Illustrated in 1999. But at the beginning of his journey, he had to slay his Goliath. We look to the culmination of his journey, the end of the story, and marvel at the results, but it's the beginning that elicits the most significant lessons.

Sonny Liston was described as a hulk with upper arms like picnic roasts, two magnificent 14-inch fists, and a scowl that he mounted for display on a round, otherwise impassive face. He was one of the most intimidating and skilled boxers of his era.

He won the heavyweight title by destroying Floyd Patterson, one of the greatest boxers of all time, with two punches, left hooks down and up, in the first round. Liston was Mike Tyson before Mike Tyson.

No one gave Ali a chance. Family and friends attempted to get him to reconsider. Boxing experts were generally concerned about whether or not he would make it out alive. The people closest to him begged and pleaded with him not to fight this fight. The people he trusted the most spoke to him with absolute certainty that he could not win this fight. Yet, against these seemingly insurmountable odds, his whisper of belief told him repeatedly:

You are the greatest!
You are the greatest!
You are the greatest!

When faced with questions and doubts about the task before him, *the voice* would get louder:

You are the greatest,
You Are The Greatest.
YOU ARE THE GREATEST. But he's so big and skilled.
YOU ARE THE GREATEST! But I haven't beaten anyone like him.
YOU ARE THE GREATEST!! Maybe everyone is right.

YOU ARE THE GREATEST!!!

The words he repeated to himself became his reality. The dominant voice within us drives us, moves us, motivates us, and ultimately propels us to success or failure. We must learn to cultivate and train this voice so that it empowers us rather than rob us of our power.

We have been subconsciously subject to a disempowering message for the better of our lives. Other forces within us fight against the message of the empowering voice. We fight against our own emotions; anger, doubt, and frustration. We fight against the self-centered voice of our ego, pride, entitlement, and justification. These entities within us drown out the whisper of the truth, constantly fighting its way to the surface.

Though most of us will never be in the ring fighting for our professional lives, as leaders, every day is a fight to bring value and lead our teams effectively. As aspiring leaders grow, they often lack the experience and knowledge to overcome the challenges our new role presents to us. To be the leader our people need us to be, we must be ready for a fight.

Often we find ourselves with the same odds or worse as Ali going against Liston. We doubt ourselves endlessly; we question every decision we make, paralysis by analysis.

Your whisper is beneath the layer of doubt, fear, and insecurity. Your whisper may not be "YOU ARE THE GREATEST," but it will be something we resonate with—we all have our whisper. That whisper may not even be your own voice, and your whisper might be your children's voice. "Mom, you are the greatest! Dad, you are the greatest!"

Their voices may be the ones that allow you to persevere and overcome. For those who serve, it may be the voice of our nation crying out to those who have been called to sacrifice their lives to protect the ideals, principles, and furtherance of freedom. It may be the voice of possibility for those working on monumental breakthroughs in medicine or science—the voices of those that will benefit from your diligence years after your existence. I'm here to feed, cultivate, and build that whisper into a roar.

Being thrust into a leadership position is one of the most daunting and overwhelming experiences one can endure. You have been thrown into a heavyweight fight, and all odds are against you. You have figured out how to get the best out of yourself through your mechanisms and idiosyncrasies. But now, the true challenge is getting the best out of the people you lead, and all the mitigations you have made for yourself to succeed will become liabilities.

When only you are responsible for your success, you don't have to answer for your attitude or an unchecked ego. When you are responsible for a team, your attitude could be the difference between success and failure.

On the surface, to the untrained eye, it may appear as though your opponents are the external challenges of leading your team to successful results. It may seem as though you're fighting against a lack of resources or not enough team talent, but your actual opponent is within you, and there is no more significant fight.

When we face new challenges, our nervous system simply attempts to keep us safe from the uncomfortable, automatically forcing us to look outward. It is painful to look within—take ownership of the problems, and accept accountability for what is going wrong.

Looking inwardly triggers the inner voice of fear, imploring us to turn around and return to what's comfortable. It triggers you to run away from the problem and blame others for the failures. Allowing that innate voice of fear to take control dismantles our ability to lead indeed and be what our people need us to be.

Leadership isn't comfortable. Leadership isn't easy, and leadership isn't safe. Leadership will present you with the greatest challenge of your life.

Initially, the voice of fear will be thunderous and clear. But this is the moment where we face the giant within. We must fight where we prepare our five smooth stones to slay the monster. We must persevere because the livelihoods of the people we have the privilege of leading are at stake. We either become the key for them to open the door to their full potential or the padlock that keeps them locked away in the prison of what could have been.

We must silence that doubt and fear that resonates so viscerally within our minds by raising the volume of belief so that we may lead fearlessly, courageously, and powerfully. The loudest soundtrack in our minds is the playlist we force on our teams, families, and colleagues.

As leaders, we have an unusual degree of power to project our shadow or our light onto other people. We play a significant role in shaping the conditions under which people live, move, and their being, conditions that can either be as illuminating as heaven or dark as hell.

Leaders must take special responsibility for what's going on inside themselves, or the act of leading will create more harm than good. The most significant factor in what we project as leaders upon our teams is cultivating what is happening within ourselves. Our inner life shapes our reality; we project that reality upon our teams.

We cast that shadow upon our teams if we subject ourselves to a disempowering voice that constantly communicates our ineptitude or incompetence, thereby facilitating an environment of uncertainty, doubt, and mistrust. We must meet the challenge within us with a willingness to fight not just for ourselves but, most importantly, for the people we lead.

The greatest leader to ever grace this earth provides the blueprint for us as leaders beginning our journey to assume the mantle of leadership.

After fasting for 40 days in the wilderness, Jesus met with the most significant leadership challenge at the very beginning of his leadership journey.

Tired and starving, the devil came to take advantage of this weakness and challenged Jesus to turn stones into bread. The devil knew that Jesus could manipulate the situation to work in his own favor, to serve his own purpose.

Oftentimes as leaders, we have the power to shift uncomfortable situations into more favorable conditions but at the risk of jeopardizing future success. We have the ability to always be correct. But is what's suitable for me right now, right for everyone I have the privilege of leading?

With unwavering commitment and understanding of what was at stake, Jesus stated that it takes more than bread to stay alive and denied himself this immediate pleasure. Jesus focused on the more important thing than his pain to fight this challenge.

As a leader, you will be frustrated, tired, and upset, and the thought of quitting will be loud in your thoughts. When you are feeling your worst, the most significant challenges present themselves. It is then that the voice of fear is screaming so loud that it is deafening.

Serve yourself, save yourself, is the cry of fear. The temptation to turn your back on the team begins to look more desirable every minute. You begin to craft a narrative that justifies your fear and tell it to yourself repeatedly.

Excuse after excuse begins to be conjured seemingly out of thin air, and submitting to your fear is merely a decision away. You must start with the end in mind and question yourself at that moment. What will be the repercussions of submitting to that voice of fear? What will happen to the team? Who will I become if I choose to serve my self-interest when challenged? Will the decision to listen to the voice of fear get me closer to the person I ultimately desire to be? We disarm the voice of fear by breaking the negative pattern through questions that help us refocus on what really matters.

Afterward, Jesus was taken up to the top of a church and told that *IF* he was the Son of God, "throw yourself down" because the angels would come and save him. But again, Jesus withstood this temptation and answered, "it is written, thou shalt not tempt the Lord Thy God" Matthew 4:7.

The devil knew that he was the Son of God, but this is what the enemy does, sow doubt so that you question your own belief. We've all had questions about our own ability, potential, and faith in ourselves. But Jesus shows us how to fight the *IF*, don't even acknowledge it. Jesus knew exactly who he was and never questioned his identity.

THE FIGHT TO LEAD

When challenged with a question about who he was, he didn't acknowledge or answer; he shifted the focus to what was most important. Whatever you focus upon increases. When we focus on what we are not, we give power and strength to our doubt, which can consume us.

When we focus on who we are, the forces seeking to detract us will never succeed. When your belief is questioned, you shift the focus off of yourself onto the purpose of your belief's pursuit.

Then for the final push, Jesus was shown everything the world could offer: all the power, riches, kingdoms, and glory that was possible. But Jesus knew all of this and more and was not tempted. Jesus responds with, "Get behind me, Satan: for it is written, Thou shalt worship the Lord thy God, and him only shalt thou serve" Matthew 4:10.

It takes courage not to seek the accolades, pursue the credit, and ensure the spotlight is dispersed on those responsible. It takes courage to serve those you have the privilege of leading.

In service, you will never be fully repaid for your efforts, you will never get paid enough, have enough time off, or get enough appreciation. Service can never be decoupled from sacrifice; they are inseparable. The reward of service is the success of those you serve.

Jesus laid the blueprint for us to be effective leaders because immediately after having his belief tested, he began his leadership journey with "Follow me." In Matthew 4:18, Jesus, unknown by Peter and Andrew, show up at their place of business and compels them to follow him with no reassurances or promises. They immediately followed despite the conditions.

This is compelling because Peter and Andrew were at their place of business doing what was necessary to feed their family. Fishing was all they knew. They had no other means of providing a life for their family, and they came immediately when Jesus compelled them to follow him.

Jesus didn't offer a more competitive salary package with a company car and a side office; he offered them the chance to catch more fish than they could ever imagine. This is the leader we have the potential to become when we choose to go to war with ourselves and eliminate the liabilities that lie within rather than focus on what we can gain externally.

Often it is at our worst times when our belief is tested. Will we do what is easy, or will we continue to embrace the pain of growth through our most challenging times?

When you become a leader, by choice or by force, the universe conspires to test your mettle. The universe conspires to run you through the gamut to see if you are capable of that title.

This is how you know you are in the position to lead. Let's reflect on the leadership in this chapter and look at what they did.

1) They had beliefs in something beyond themselves.

2) They never quit, no matter the odds, because their conviction was stronger than the pain of rejection.

3) They took time to their craft, learned how to fight, and learned not to listen to the inner voice. The belief was more substantial than any power punch.

Even the Savior of the World gave us a powerful example of not falling into temptation, and he pulled his strength from God and fought off the very devil himself.

Think about your leadership skills. Think about your strengths and weaknesses—in the examples in this chapter, which one resonated with you? Ask yourself why that is—jot down notes because, like any great fight, we have many more rounds to take on, lessons to learn, skills to practice, and awareness that we will have a lifetime of practice ahead of us.

What will round two reveal to you? That is exactly what it needs to show you. Now, let's punch on through to the next chapter.

ROUND 2

THE CHALLENGER—EGO

Don't criticize them; they are just what we would be under similar circumstances.

—ABRAHAM LINCOLN

Many historians believe that President Lincoln's greatest victory occurred after the Civil War or the abolishment of slavery. I contend that his most significant victory is the one he achieved over himself.

In the summer of 1842, a young Abraham Lincoln was practicing law in Springfield, Illinois, and he also supported a political party known as the Whigs. The country was going through a financial depression under president Andrew Jackson. People did not trust the banks, wage growth was stagnant, unemployment was at a record high, and they didn't trust the government—sounds familiar.

In 1842, the state bank of Illinois was on the brink of bankruptcy, and families lost their entire life savings. At this time, there was no unified American currency, and individual banks printed their own banknotes backed by silver and gold. Illinois chose another Springfield lawyer to handle this crisis, James Shields.

Shields held the position of the state auditor, where he was responsible for managing the budget and balancing his loyalty to the democratic party. Shields and the state governor signed a proclamation ordering individuals to pay their taxes in silver or gold rather than banknotes.

Lincoln disapproved of the proclamation because he believed it would cripple lower-income individuals and farmers. Lincoln and the Whigs thought the proclamation was a selfish order to put state official salaries over the welfare of the *common people* they were chosen to serve.

Lincoln was furious and decided to take action. Politics then were very similar to the politics of today—mud-slinging was the go-to strategy. The difference then was there was no Twitter to leverage the choice words one would have for their opponent that the world could read in an instant.

In the first known use of a burner account to spread accusations and insults about their opponent, Lincoln wrote several anonymous letters to the editor of the Springfield Journal, a local Whig newspaper. Using the moniker of Aunt Rebecca, Lincoln referred to Shields as a fool, a liar, and a conceited dunce over several letters.

Lincoln decimated Shields's character in those letters causing the public to further question the integrity of him and the Democratic Party. Shields and his team were incensed and demanded from the newspaper the true name of Aunt Rebecca. Being pressured fiercely by Shields, the editor gave up Lincoln's name to Shields.

Lincoln was terrified at the thought of a duel with Shields. With his reputation and integrity damaged, he was forced into only one option; challenge Lincoln into a duel. Shields was a fantastic shot, and Lincoln realized he had bitten off more than he could chew.

Lincoln was known as a quick-tempered and quick-witted young man who, up until this point, never had to answer for his irrational outbursts of anger. His lack of maturity, insight and emotional intelligence led him to a situation that could have cost him his life. Lincoln would obviously go on to be president, but there was a real possibility of him dying at the hands of Shields.

The future implications were gargantuan. Our future as a nation would have looked significantly grimmer without the leadership of President Lincoln. Who knows if the enslaved people would have been freed? Who knows if the elected president would have been able to navigate the rise of the confederacy? Would there have been the United States without the leadership of Lincoln?

Why? All because of the lack of internal self-management of his emotions and impulses. Individuals can not approach the mantle of leadership until their inward life of that individual is appropriately managed.

Though most duels were fought with pistols, Shields was an amazing shot. Luckily, the person challenged to the duel can decide the rules of engagement. Lincoln leveraged his massive size advantage and chose cavalry broadswords of the largest size. Lincoln had a seven-inch reach advantage with his 6'4" frame versus Shields's 5' 9" frame.

On the day of the duel on a sandbar within the Mississippi river, local citizens gathered around the banks to watch the two men prepare to fight. Shields tested his sword by slashing through the open air while his second begged for him to call off the duel. Shields declined to quit while watching Lincoln slice down a large tree branch with his sword on the other side of the bank.

As the two men approached each other for the duel, Colonel John Jay Hardin jumped in and stopped the proceedings within seconds.

Face to face, Lincoln explained to Shields why he did what he did and subsequently apologized for his actions. James Shields accepted Lincoln's apology, the broad swords were sheaved, and the two men shook hands. Lincoln, from that moment, vowed that he would never write another hurtful word or speak ill of another person. Vows have a way of being tested along our journey of life.

Fast forward twenty-one years, Robert E. Lee of the Confederacy was determined to invade the Northern Territory to capture Washington. The Union and the Confederates would meet at a small Pennsylvania town famously known as Gettysburg, where they would fight the most famous battle in the history of the United States.

During the first two days of the fighting, the Union army lost nearly 20,000 men. Lee was eager to wipe out the remaining Union forces by a tactic out in the open led by General George Pickett. This attack, known as Pickett's Charge, became the most catastrophic charge ever known in the western world. Pickett's brigade commanders, except for one, were killed in just a few minutes, and 4,000 of his 5,000 men had fallen.

In the late hours of July 4th, 1863, Lee, with his remaining men, began to retreat from Gettysburg. As they were retreating, heavy rainfall began making the retreat impossible. Leading the Union during Gettysburg was General George Meade.

History paints General Meade with a narcissistic brush, but his experience in military affairs was long and valued. Once the news of the Confederacy's positioning reached Lincoln, he was ecstatic—an opportunity to pursue and overwhelm the South and possibly end this war very soon.

Lincoln ordered Meade to attack repeatedly, but General Meade did nothing. He stalled with caution disobeying the orders of the President. This allowed the waters to recede and the Confederate Army to escape, prolonging the war.

Furious, Lincoln began to revisit the errors of his youth and put pen to paper writing:

> My Dear General, I do not believe you appreciate the magnitude of the misfortune involved in Lee's escape. He was within our grasp, and to have closed upon him would have ended the war as it is.
>
> The war would be prolonged indefinitely. Your golden opportunity is gone, and I am distressed immeasurably by it.

Lincoln re-read his letter and thought about the situation at hand. Historians have theorized that he most likely concluded that this letter would have personally hurt the general. So Lincoln put it aside. The letter was only seen again in the 20th century when the president's documents were reopened. Underneath the letter to General Meade was a notation stating, never sent and never signed.

Lincoln truly learned his lesson from the *near duel* in 1842. He displayed sympathy rather than criticism and restraint rather than recklessness. Lincoln wrote many letters of criticism along the way, but they always ended up the same: never sent, never signed.

Lincoln was a master of the inner game. Lincoln could have easily been encumbered by the frustration of his generals' failures. He could have criticized and blamed and been absolutely justified in this approach.

When things go wrong, we often seek justification rather than ownership. Within us, our ego tells us all the things going wrong are the faults of others as a means to protect ourselves. The need to protect ourselves locks our perspective in the past, and we cannot solve the problems that are still present.

Lincoln acknowledged his frustration and anger by writing it down in a letter but took complete ownership of the problem by never sending it. This allowed him to remain calm and clear and see things properly.

As a result, he could find the general the Union Army truly needed. Lincoln selected Ulysses S Grant as Lieutenant General of the U.S. Army and was answerable only to Lincoln.

Grant was energetic and strategic, and he earned the trust of the President through his actions and had the immense respect of his troops and civilians alike. Grant's relentlessness and unwillingness to accept defeat led to the South's surrender in 1865.

Often our ego betrays what is prudent and reasonable and causes us to act irrationally. Lincoln's fight was never with General Meade or James Shields; it was always with himself, and his own ego was his enemy.

Lincoln's weapon of choice to defeat his ego was the pen. Anytime Lincoln received information that was problematic to his cause, he would immediately put pen to paper and write his frustrations and displeasure with the person or situation. This allowed him to express his displeasure in a healthy and non-consequential manner.

Lincoln never allowed himself to see people in situations through the emotions generated by a bruised ego. When Lincoln released those emotions on paper, he could see the world as it indeed was and not with him as the center.

When we see the world through the emotions of a fractured ego, we can only see how we are affected. We become severely self-centered and are incapable of leading when our focus is solely on ourselves.

As leaders, this fight with our ego is a daily struggle. When our people aren't performing to the level we believe they should be performing, we feel the pressure of the organization because it reflects our capabilities. This pressure ignites our ego, and we risk becoming resentful of the people we have the privilege of leading. This resentment creates an invisible barrier between the leader and the individual, preventing the leader from providing them with what they truly need to succeed.

Do what Lincoln did when you feel like you're fighting your ego. Get your pen and paper out and disarm the power of your ego so that you may continue to lead with clarity.

Due to how our brain processes information, writing down our frustrations is backed by scientific evidence. We have two hemispheres in our brains. The right hemisphere is our creative side that processes information more abstractly. This is where art, music, emotion, and our feelings are processed. The left hemisphere is our logical side. Math and language are processed on this side, and the information flow is analytical and structured.

When we are deep in our emotions or our ego has taken over the messaging and meaning, we operate primarily from the right hemisphere. If we allow the moment's emotions to overtake us, things can escalate negatively very quickly.

This is one of the primary reasons you have heard the advice of counting to ten when you're upset. It isn't the counting that is an antidote to the negative response; it's the left-brain interaction that allows you to see through a more logical and reasonable lens because you are counting.

We also disable the emotional response from our right brain by conducting a left-brain task of writing down our emotions. Writing out emotions helps you prioritize problems, fears, and concerns, allowing you to recognize triggers and learn ways to control them better.

Lincoln did this as his pen fought through the emotions of his mind. With his placement of the letter and stroke of his pen, the words left became a magnificent reminder that the pen is mightier than the sword.

Round two challenges us to control our Ego. Lincoln's example of using words first to pick a fight or a duel with Shields is a huge social media lesson for us.

How many emails have you wished you could have *unsent*? How many times did you wish the return key on your keyboard meant just that, return to sender and not cause pain and later embarrassment?

What can you learn from Lincoln? Will you do as Lincoln did and write it out so that your emotions are satisfied and then rip, shred, or delete it? Here is something to remember, "When in doubt, delete it."—Chellie Phillips. Now, let's discover the lesson in round three.

ROUND 3

THE CHALLENGER—PAST HURTS

The most essential factor is persistence—the
determination never to allow your energy or enthusiasm
to be dampened by the discouragement that must
inevitably come.

—JAMES RILEY

"Your type doesn't last long here; I'm not going to waste my time getting to know you; I'm done with you!" This was my first week on a submarine, 18 years old, and this is the motivating speech I received from my Executive Officer (XO), second in command. This shot to my gut lingered for a long time, and it took me a while to get back to my feet. The opponent for this round is past hurts.

Recently, my two daughters, Justice and Faith, have been taking piano lessons. With the onset of the pandemic, we have been relegated to our homes, and they are forced to conduct their lessons over Zoom with their instructor. Luckily, we have a piano at home that we were gifted several years ago. To be honest, we mostly had it for decoration, but now we have been forced to utilize it in real life for their lessons.

We contacted several local professional piano tuners to tune our severely out-of-pitch piano. Surprisingly, the tuners quoted us in the thousands for our piano because it was so out of tune. They informed us that the strings are powerful; to keep them at the perfect level of tension, we've got to condition them to stay at this level.

Notes must be tuned, so they are accurate and stable. It is one thing to get a note to the desired pitch; it is another for it to keep its pitch. Piano strings hold much more tension than guitar strings, so stabilizing the pitch takes more nuance than just turning the tuning pin.

Getting a note in tune is just the first step. Professional tuners strike the same note repeatedly at a loud volume to ensure the notes will hold. Fine tuners utilize many techniques to ensure the notes will hold beyond the first hard blow.

Every string has a tremendous amount of tension, and setting the strings properly is a challenging task. Strings have what is called speaking length and non-speaking length areas to them. The speaking length is the main vibrating portion of each string that provides the basic tone when the hammer strikes the string.

The non-speaking lengths are at each end of the string and are the little sections between friction points that are not usually part of the sound when played. In between these sections, you have points of tension. Setting a string requires you to deal with the tension points and how they will even out once you strike the string.

My wife was unwilling to pay the enormous price for the tuning, especially since they would have to return and re-tighten them regularly until the wire was trained to stay at the proper level. In their current state, the keys produced a sound inconsistent with their purpose.

Armed with the tools of champions, Amazon, and YouTube, my brilliant and ambitious wife tackled the daunting task of tuning the piano.

Immediately after she completed the tuning, the keys would sound phenomenal. After only a few days, the keys would need to be readjusted again. This process was repeated over a couple of weeks. Over and over again, things would sound great, and then after an undetermined amount of time, they would drift back out of pitch.

During this process, I thought about the similarity in the struggle it was for me to get myself *back in tune*. So much of my early leadership journey mirrored a *Jeckyl-and-Hyde-type* relationship. One moment I would do all the things I learned from mentors and all the leadership information I read and absorbed. The next moment I would snap back into old habits of yelling and demeaning behavior, limiting the potential of my team.

At the root of my anger and constant outbursts resided the pain and pressure of believing I needed to prove myself. I believed every mistake my team or I made was a direct indictment of me. This made me feel unworthy of the position unless things were going absolutely perfectly.

When I showed up on my first submarine, I had to do intake interviews with the Commanding Officer, Executive Officer (XO), and the Chief of the Boat or COB. These interviews were a means to learn about the new crew member, background, and goals.

The leadership team would also communicate their expectations to the new crew member so they could clearly conduct business on the submarine. I showed up extremely excited about joining this elite team of submariners.

I met with the Chief of the Boat, and he was an intimidating figure that made his expectations clear for how I was to conduct myself on his submarine. He explained the importance of becoming a person of value onboard.

Submarines are limited to around 130 crew members, and everyone has a vital role in ensuring that the submarine can complete its mission and remain safe. Purposely submerging a black tube full of people is a dangerous game, and the weight of that reality became clear during our interview.

I was fired up and ready for the challenge of contributing to the crew of this great boat until my interview with the XO was not as uplifting, to say the least. I walked into his stateroom, and he asked me for my record without even looking up at me.

I stood awkwardly while he rifled through the chronicles of my short career. As I went to sit down, he motioned for me to continue to stand and sat back in his chair, finally looking at me with obvious disdain. He said, "You've been in the Navy for two minutes, and you have already been in trouble. You've been given this awesome opportunity to do something with your life, and you're gonna squander it away."

He then said the words that would live rent-free in my head for the next ten years of my career, "Your type won't last long here. I'm not going to waste my time getting to know you. You can leave."

As I exited his stateroom, those words, *Your type won't last long here*, did a continuous loop in my head. At first, I was in shock and disbelief. I couldn't believe that he would say that to me.

My shock and disbelief rapidly turned into sadness and then anger—so much anger. I resented him deeply, and that resentment scarred me. I was severely out of tune. The pain and self-doubt that were generated from that moment fought against my willingness to be the leader I knew I needed to be.

Your type won't last long here. What type was I, and why won't I last? These uncertainties became embedded in my psyche, shaping the lenses through which I saw everything on the submarine. I didn't feel part of the crew because of my *type.*

Operating a submarine successfully is akin to an orchestra; everyone must play their part and conduct their role perfectly for our mission to succeed.

I was not invited to play in that orchestra and merely existed rather than contributed for a long time. I didn't seek to collaborate because all I heard in my head was, *your type won't last long here.* My psychological posture at all times was survival.

Everything that I did was based on self-preservation. I worked hard and received accolades along the way, but they were merely a byproduct of trying to survive in an environment where I was neither given a map nor a compass.

The XO's voice drowned out any whisper of success. I was severely out of tune, and it took a herculean effort to get me back in tune.

Just like the tuning process on the piano, our inner voice is our own personal instrument. If left unprotected or underdeveloped, our inner voice begins to drift far away from its purpose and subsequently drives us from our purpose.

It begins to be influenced by the environment. Just like the keys are affected by the humidity, the temperature, the constant moving and jerking, the dust, and systematic abuse from random 4-year old's banging on them. Our inner voices are negatively affected by the environment and the media projecting its biases and generalizations upon it.

Friends and family project their insecurities and limitations upon it. In my case, a person with enormous influence in my chosen profession negatively affected my inner voice by launching his own fears upon me. From this projection, everything confirmed his beliefs, and my failures provided evidence of the inevitability of future failure.

I believed his opinion because of who he was, and my whisper of success began to diminish daily. I never even considered the reality he was speaking from; I accepted it as my reality.

When someone tells you that *you can't* in terms of your ability, they are communicating, "In my version of reality, there does not exist a possibility of this occurring or ever existing."

When someone tells you that you can't achieve something, it really has nothing to do with you but everything to do with their limitations.

I had to divorce myself from his version of my reality to get back in tune. This wasn't something that I could accomplish alone. During a professional fight, it is one versus one. You're in the ring alone against your opponent. In this fight against yourself, you're not only allowed to bring in assistance; it is encouraged.

To defeat this battle against these past hurts, you need to bring someone to this fight with you—a trusted agent, a mentor, a coach, a friend, anyone who knows the truth about you. For me, it was a Senior Chief named Ty.

At the time, he didn't know he was my mentor, but he was the most knowledgeable and influential person onboard the submarine.

I gravitated toward him because of his presence and professionalism.

He embodied what I wanted to be as a Sailor. He wasn't the typical mentor with motivational words of encouragement and inspirational quotes. Not even close; he was tough on me. He was highly demanding and borderline mean. He never let up on me or let me take the easy way out. There were times when I was absolutely convinced that he hated my guts, but one moment changed everything for me.

We had the privilege of pulling our submarine into Roosevelt Roads, Puerto Rico, for a port call. My brother and his wife were stationed in Puerto Rico, so the leadership allowed me to take leave and spend time with them.

When a submarine pulls into town, an invisible magnetic force attracts all the submariners to the same bar. It's an unspoken thing that happens, but my brother and I decided to go out and have a few drinks, and low and behold, we wind up at the same bar as the rest of the submarine. I was excited to see most of the crew so I could introduce them to my brother.

As I was going around the bar introducing my brother, I saw Ty coming my way out of the corner of my eye. I was a little nervous because I had no idea what he would say to my brother about me. I told my brother who Ty was; he probably would get on me a little because he was always hard on me.

To my surprise, as I introduced my brother to Ty, he interjected and told my brother that I was one of the finest Sailors aboard the submarine. He said to him that I was brilliant, capable, and had a bright future ahead of me. He smirked as he directed his words toward me.

"Armon, I'm hard on you because I believe you can be great. You are the top E-4 onboard at your best but at your worst, whew. My goal is to get you performing your best at all times so that everyone can see what I see."

Afterward, Ty returned to the bar to get a drink, and my mouth was wide open in shock. My brother talked to me, but I didn't hear a word. This person I thought believed the worst of me believed that I could be great.

From that moment, the declaration from the XO that my type wouldn't last long began to diminish, and the words from Ty became the most prominent voice in my head. His words repeated in my head over and over again, "...I believe you can be great." How did someone so influential and prominent in my organization believe this about me?

The voice of encouragement from Ty silenced the voice of the XO, but it was temporary. It was the first time I felt in tune in over a year. I wasn't angry or frustrated, and I was performing the best I had in months.

Every now and then, the XO's words would creep in, but adding one step in the process allowed me to silence his voice for good. Sometimes, when we receive encouragement from our mentors, we have to believe in the belief that they have for us until the belief that we have for ourselves catches up.

During this round, our opponent is the past hurt that keeps us from believing in ourselves, keeps us bound to negativity, and keeps us from performing to our potential. As a leader, those past hurts allow doubt to creep into our minds about whether or not we are even qualified to lead. It's those past hurts that cause leaders to experience imposter syndrome.

Impostor syndrome is a psychological phenomenon in which an individual doubts their abilities and accomplishments and fears being exposed as a *fraud*. It often occurs despite our level of competence, leading to feelings of inadequacy and self-doubt.

Imposter syndrome causes leaders to constantly seek validation and reassurance, which can lead to micromanagement or failure to act because of the constant need for feedback. It can also lead to overworking and overcompensating, leading to burnout, therefore, causing emotional outbursts in response to challenges.

To win this fight and get ourselves back in tune, we must bring someone into the battle with us who can nullify the negative effects of those hurts and remind us of the truth about ourselves.

We need those mentors to help along the way to provide an objective perspective to help us reframe those negative thoughts and provide a clearer and more accurate view of our abilities and accomplishments. We need those mentors to let us know when we are out of tune and help us reset ourselves. Who is that person for you? Who will be your Ty? How will you use them to help you?

All of us need great mentors to become the leader we must be. We must be willing to get past our past, heal from our hurts, and understand that people's lives are at stake based on our leadership. At the end of the day, it falls to you to make the call, do what's right, be humbled when you are wrong, and stand for your convictions and not for your past hurts. It's a difficult fight to win, but with the help of those around you that care for you, this is another opponent that can be put down for the count.

Don't fear, don't worry—just do. Speaking of fear, chapter four will talk specifically about that place of uncertainty and discomfort because of fear. Leaders, when this happens, take notes, face it, and let's help you to break down those fears and lead forward.

ROUND 4

THE CHALLENGER—FEAR

The way we communicate with ourselves ultimately determines the quality of our lives.

—ANTHONY ROBBINS

As if in slow motion, LT Colonel Chamberlain watched in horror as the body of one of his young soldiers was violently thrown back from a Confederate round not more than two feet from him. What would you do if you witnessed this?

Stepping into the ring with fear is one of the greatest fights you will ever face as a leader, but it's a fight you can and must win. We never know the future implications of our ability to defeat our fear.

I'm sure that when Joshua Chamberlain decided to become a college professor, he never thought he would be thrust into a leadership position that would determine the very fate of our nation. We never know the day or the hour when the necessity of a courageous leader is required, but as a leader, we must be ready for that day.

Joshua Chamberlain was a college professor from Maine who volunteered during the Civil War with the Union Army. Chamberlain built a reputation during his time as a professor who would not succumb to the majority opinions of those around him.

As the war broke out, Chamberlain was pulled into service based on his belief in preserving the union and his moral conviction against slavery. His father and family were vehemently opposed to him serving in the war mainly because they knew that his brothers would follow suit once he enlisted.

Sitting down with his father to reveal his decision to enlist was one of the hardest things that Joshua had to do. Joshua saw his father as a man of great character and integrity, and gaining his approval to enlist was very important to him. His father challenged him to recognize what he was doing to his family with this desire. Had he considered the chances of leaving his wife a widow and child fatherless?

Joshua stood firm on his convictions about his necessity to serve and ultimately received his father's blessing to enter the war. Even during his teachings, he would routinely speak to his classes on the importance of serving in the war. "I fear this war, so costly in blood and treasure, will not cease until men of the North are willing to leave good positions and sacrifice the dearest personal interest, to rescue our country from desolation and defend the national existence against treachery." The resonance of Chamberlain's inner voice was a combination of two words: duty and service.

Often, we desire to choose the *road* less traveled to be leaders. The *roads* that are the closest to us, that love us and implore us, we follow them to determine the easy route. They lovingly project their fear upon us, not to stop our growth but because they don't want the pain to come to us.

Choosing to lead is running to the hard thing. Choosing to lead recognizes that the outcome may not be favorable for you but gives the people you lead the chance to succeed. Chamberlain made his decision to serve despite everyone he loved protesting his decision.

We never know where our leadership journey will take us. More importantly, we never know how crucial our leadership will be for a particular circumstance. Chamberlain did not have the best training, mentorship, or environment to prepare him for what was to come in the small town of Gettysburg.

Chamberlain, the scholar-turned soldier, took advantage of his position as second-in-command and studied "every military work [he could] find." Chamberlain was under the tutelage of his commander, Col. Adelbert Ames, a West Point graduate.

Chamberlain studied, read, asked questions, became informed, and did not sit in fear but took action. When we decide to be a leader worth following, we must be aware of the challenges, obstacles, and rough terrain that must be traversed. But it is our responsibility to meet the uncertainty and fear of the moment with resolve and determination.

Upon enlisting, Chamberlain quickly found himself amid some of the war's most significant battles: Antietam, Fredericksburg, and Gettysburg.

The Confederate Army, under the lead of General Lee, was winning battle after battle. This forced Chamberlain to get up to speed quickly. Leaders, has this ever happened to you? You find yourself learning, and you're leading? What are your options? What do you do? Do you fall back on bad habits or enact your skill set and improve the situation? Falling back into bad habits is our go-to unless we have conscious thought to implement the skills we have. This is what Chamberlain did time and time again.

He out-strategized and out-maneuvered the Union army time after time. They were pushing deeply into Union territory, and if things didn't change quickly, the Confederate Army was poised to end this conflict with them as the victors.

The Union Army needed leadership. Lincoln fired multiple Generals for being timid and fearful. Desertion became a significant problem for the Union Army. Chamberlain went on to serve as the Lieutenant Colonel of the 20th Maine regiment.

He was offered to command his regiment initially, but he declined because of his lack of experience. There was no time to learn the ropes as he was almost in battle once he arrived on the 20th. After several near-death experiences during his first skirmish, he quickly learned the best way to serve his unit. He quickly earned the respect of his men through his resolve and the care he showed to his troops.

During quite possibly the lowest point of the 20th, a smallpox outbreak ravaged their regiment. The 20th was discouraged, dismayed, hungry, and confused about their current direction. It was one thing after another, and even when they thought things couldn't get any worse, this disease began to decimate them further.

Chamberlain had shown himself worthy of the mantle through his performance and leadership in previous battles. The smallpox epidemic caused the 20th to be separated from the rest of the Union Army. It forced Col Ames, a top battlefield strategist, to leave command to Chamberlain because he was more valuable in the fight.

Chamberlain had been the trusted second to Col Ames, and Chamberlain was very comfortable in that role. He was content in the supporting role, behind the scenes making sure that the 20th was ready for the battles yet to come.

However, some moments occur when you can no longer be comfortable on the sidelines. Some circumstances require true leaders to step up even though they are still being prepared.

Michael Shaara wrote a novel called *Killer Angels*, where he wrote a historical account of the Battle of Gettysburg extracted from the journals of those who fought in the battle. There was a minor but ensuing incident prior to going into battle. Chamberlain was given custody of 120 Union mutineer prisoners just before the battle of Little Round Top.

Chamberlain was granted authority to shoot any prisoner who refused to follow his orders. But that is not his way. Rather than threaten these soldiers, he took a different approach. He looked at these forces of 120 men that would mean the difference—he needed them to strengthen his regiment for the upcoming battle.

It is recorded from the book that a mutineer, a prisoner, tells Chamberlain, "They been tryin' to break us by not feedin' us." With that information, Chamberlain gave them food and water.

Next, Private Bucklin from the mutineer group approaches Chamberlain and said, "...He was selected to tell him of the prisoners' grievances." Chamberlain listened intently to Bucklin rather than ignoring him. Chamberlain had a choice: take the time to listen to Bucklin or ignore the requests and hope they will join and strengthen his regiment.

Chamberlain said this in response:

> I've been talking with Private Bucklin. He's told me about your problem. There's nothing I can do today. We'll be moving out in a few minutes; we'll be moving all day.

I've been ordered to take you, men, with me. I've been told that if you don't come, I can shoot you. Well, you know I won't do that. Maybe someone else will, but I won't.

The whole reb army is up the road a ways waiting for us… We can surely use you, fellows. We're well below half-strength, and whether you fight or not, that's up to you.

Chamberlain continued:

You know who we are and what we're doing here…This Regiment was formed last summer in Maine. There were a thousand of us then, and there are less than 300 of us now. … [We came for different reasons.] Many of us came because it was the right thing to do.

This is a different kind of army. If you look at history, you'll see [that] men fight for pay … or some other kind of loot. … But we're here for something new. … We're an army out to set other men free. …

With this final statement, Chamberlain laid out the entire decision.

If you choose to join us if you want your muskets back, nothing more will be said. If you don't join us, you'll come along under guard. When this is over, I'll do what I can to see that you get fair treatment. … Gentlemen, I think if we lose this fight, we lose the war.

What were the key points he said to these men?

1. He treated them with human dignity by feeding them.
2. He listened to them.

3. He spoke to them respectively, truthfully, and gave them a purpose. What was the result? Chamberlain gained their trust through these actions. All but six of the 120 mutineers chose to join him.

The battle of Gettysburg was the turning point of the Civil War, and Chamberlain was a key factor in it. The battle had been raging for two days, with the Confederates recognizing the vulnerability of the Union in this position and heavily attacking the left flank to gain a significant advantage in the battle and the war.

It was imperative to hold their flank position at Little Round Top against attacks by the Confederate Alabama 15th Regiment. The flank position in a regiment is the right or left edge of the regiment that is not oriented toward the enemy. It is typically the weakest area of the regiment because the strength is concentrated towards the middle. If the enemy takes the flank, they can surround the regiment and defeat them from the inside out.

His commanding officer told Chamberlain, "You cannot withdraw under any conditions. If you go, the line is flanked, and they'll go right up the hilltop and … [attack us from the rear]. You must defend this place to the last."

A Union loss at *Little Round Top* could have changed the outcome of the *Battle of Gettysburg*, the Civil War, and perhaps the course of history. Chamberlain's Regiment held the high ground, giving it an advantage over the Confederates, but they had suffered heavy losses and ran out of ammunition, risking being overrun.

Wave after wave, the confederates sent men trying to take that left flank. Chamberlain remained steadfast as his men died, firepower diminishing, the enemy gaining more ground; he just kept repeating:

"We will not turn back; we will not fail; we will not retreat!!"

"…We've lost a third of our men. "WE WILL NEVER RETREAT!"

"…We have no more ammunition. "WE WILL NEVER RETREAT!"

"…Sir, we have to pull out! "WE WILL NEVER RETREAT!"

"…Sir, we can't hold them again!" "WE CAN AND WE WILL!"

Looking down the line, Chamberlain knew that if the line broke, they would lose the flank, and the Union Army would be decimated. NO! Going back was not an option; staying where they were with no ammunition meant certain death. NO! The call came out, "Fix bayonets!"

When Chamberlain gave the order, no one moved. His regiment looked on in complete confusion. His idea, birthed from a resolved foundation, didn't resonate immediately with his soldiers. They believed giving up was an option, so they were stunned when Chamberlain presented the idea of attaching the bayonets to the end of their weapons and charging the enemy. He repeated the order to fix bayonets and, on his order, charged down the hill toward the confederates.

Chamberlain and the 20th were severely outmanned five to one. They were entirely out of ammunition, and there was no option to retreat, leave, or give up. This decision was as close to certain death as it gets.

But death would not receive Chamberlain and the 20th on that day. With bayonets fixed, consumed with rage and resolve. They charged down the hill screaming, roaring, and rising above a sure demise. Overcome with bewilderment, confusion, and fear, the confederates stopped their advance, froze in place, crouched down, and then reversed their course. They turned and ran. They threw their loaded muskets to the ground and ran in fear from the advancing 20th.

Chamberlain would say about that day, "But the cause for which we fought was higher, our thought wider...That thought was our power." The thought was his strength. The constant revisiting of this thought, this higher ideal, was what he repeated to himself in the moments before.

Chamberlain was a professor, a man of education and languages, but not the sword. He wasn't born a warrior; he wasn't made that way. But through study, experience, and facing the battle head-on, he made himself that way; he made himself into what was needed for that situation by communicating with himself that he would be the factor that kept truth and the Union strong to win it.

Chamberlain was a visible leader who led from the front. He communicated to his men the importance of the mission and their role in fulfilling it. His men respected and trusted him because Chamberlain respected and trusted them, and they gave Chamberlain their loyalty and maximum effort.

Who or what is your Gettysburg as a leader? What will you say to yourself during your moment of truth? What will you say when caught between a rock and a harder place? What will you say to yourself when everyone else says to give up?

Will you lead the charge with your modern-day bayonets fixed and ready to charge into the fray? How will you keep your positive self-talk going when all seems lost?

Chamberlain communicated with himself in a way that allowed him to only think about options to win. He spoke with himself in a way that ignited an overwhelming belief in his actions. At that moment, he believed in his actions so much so that not only was he convinced, but the confederates were also convinced that they wanted no part of this bayonet-crazed professor charging at them with all he had.

Nothing can withstand the power of the human will if it is willing to stake its existence to the extent of its purpose. In this fight against overwhelming odds and the voices of defeat, the leader shows themselves strong.

Through this example, what have you learned about leadership? What did Chamberlin do in facing all odds? Why did he make the decision he did?

After pondering on these questions, I want to ask you if you remember the title of this chapter: The Challenger—Fear. Fear has the influence to make us stop just at the moment of victory. Fear makes us want to run, flee, and shut down. It is a massive trigger for our brain to switch to fight or flight.

However, if you are a leader—especially in a battle situation—you must fight that survival response and make the right decision. It is forever debated on what could be right or wrong in any given situation, but that usually is in the aftermath in 20/20 hindsight.

When you are in the moment, you must rely on your training and your strength and look at the best situation based on the knowledge that you have at that time. During a tour or an assignment, those few moments will reveal our true mettle as the leader. The voices of defeat will be deafening during those times. Who will you choose to listen to? What will you do? Allow your voice of triumph to rise up, fix bayonets and take that hill.

ROUND 5

THE CHALLENGER—THE ENVIRONMENT

*I learned I was not, as most Africans believed, the victim
of my circumstances but the master of them.*

—LEGSON KAYIRA

9,702 miles.

Traveling 9,702 miles by car would take approximately 153 hours of non-stop driving. By comparison, it is more than 38 times the distance between London and Moscow or more than Thirty-nine times the distance between Paris and Moscow. It's also longer than the distance between the two most distant points on Earth, the antipodes of each other, which are about 8,000 miles apart. It's a long distance to travel by any means, and it illustrates the determination, grit, and resilience that someone would need to undertake such a journey.

What drives a man who was left for dead at birth, with parents illiterate and unable to care for him properly, to decide to look into the abyss of his reality and see the light of opportunity?

What truly makes a leader? A leader is someone who is accustomed to overcoming difficulty. A leader is someone who is comfortable in adversity because they have navigated the depths of difficulty successfully before. There are very few things that impose adversity and difficulty on the leader than the environment. Our environment is an unknown quantity that we have very little control over.

We can't choose our parents. We can't choose our family. We can't choose the location where we grew up. We just have to deal with what's presented to us. More often than not, we conform to the social norms of the environment we live in. Everything within our personal sphere influences how we see, how we think, what we believe. Our environment shapes our world-view. Typically when we experience things outside of our world-view we naturally reject the premise.

As a young man, in my world-view, the only type of music that was acceptable was jazz, R&B and hip-hop. I was so rigid about my belief that to me, it was a fact. It wasn't until a good friend of mine Jason Yuengling introduced me to Metallica One, that my eyes were opened. Now, my world-view has been filled with new and different genres of music that I wouldn't have naturally gravitated towards.

Even though this is a bit tongue and cheek, the idea is the same; unless something is a part of our worldview, we won't naturally consider it to be a possibility for us. Our environment is a primary determinant of our ultimate level of success and as a result could limit the level of success of our people. When we allow our environment to dictate how we proceed, it places a lid on our team's potential.

How was the environment you grew up in? According to Julian Barling, the impact of the environment looks like this. "The effects of poverty on children's schooling is a direct hit to their leadership emergence for several reasons," the researchers note. "First, general intelligence is one of the strongest links to leadership emergence, and growing up in poverty significantly affects children's academic development. But more importantly, perceived intelligence is an even stronger predictor of leadership emergence."

Think about it. If you are at the bottom of the opportunity barrel (so to speak), who will use positive influence and power for you? The statistics say that less than 3% can rise out of their environment with the proper education, energy, vision, and knowledge to execute the right actions to improve their lives.

In spite of the science, turns out, this bout is a winnable fight. Though an overwhelming opponent, we do not need to surrender to the constraints of the environment. A phenomenal man with a phenomenal story gives the blueprint for overcoming the environment. Legson Kayira's story of overcoming his environment shows us all how to do so. In a remote village in Africa, a giant baby was born. Soon after his birth, unable to feed him, his mother threw him into the Didimu River. Legson Kayira's story sounds as though it was forged on the pages of the Old Testament.

But there was no basket he was placed in, and she didn't stay to see his fate; she left immediately, unable to watch her child die. Luckily, a passerby saw him and saved him.

Legson was born in a small village called Mpale in Malawi, under extremely impoverished circumstances. He was a larger-than-average baby, and his mother was too ill to care for him. After being saved, he was returned to his mother, who was admonished severely by the village.

In his book *I Will Try*, Legson says, "I came from one of the poorest families that God ever created since the beginning of time." From the time he could walk, he was on the farm, and as soon as he could hold a spear, he was hunting. With the proceeds from an entire harvest, his family could pay the .87 cents for Legson to go to school.

The school's motto, "I Will Try," deeply inspired the gifted Legson. Contained within those three words is how this fight is won. What is trying? Is it as simple as continuing to put forth the effort to succeed at some task, activity, endeavor, etc.? Try can easily be replaced with I will strive, I will struggle, I will seek. So much in a simple word. Yoda famously pronounced in Star Wars, "Do or do not, there is no try." This is a great quote in theory but not execution. For an action to take place there must first be present within the person the willingness to step out without knowing if it will succeed. This is the very definition of trying. Without trying, there is no doing.

How does trying work in leadership? You put effort into trying to understand how to engage, motivate and manage people. It is not about perfection, it's about effort. It also means that you never quit but adjust, rethink, and try, try again. Legson did this every day of his life.

I remember stories from my parents attempting to emphasize the importance of education that would contend they walked 50 miles to school every day… uphill… both ways… in the snow. Legion's early dedication to attaining an education was the real-life version of my parent's exaggerations. He walked 16 miles every day to school with no shoes. 16 miles… every single day.

Why would anyone do that for their education? No one else in his environment was sacrificing themselves in this way. What drove him with this overwhelming force?

What is the value of education for you? Did you value yours? Think about your High School years and ask yourself the question, "would I have walked 16 miles every day so I could learn?" Getting an education is a normal aspect of our existence here in America. Because it is a normal aspect of our worldview, we have a tendency of taking it for granted and therefore it doesn't have as much value. His education was the spark that ignited the driving force to try, despite his conditions. Legson extracted every bit of value out of the education that was provided for him. For him it wasn't about just school, this was the key to advancing his life further than anyone had ever seen.

What was the last book you read, the last course you took, or better yet, do you have a coach to better yourself? What are you doing to improve yourself? Wait! Did I catch you off guard by any of that? Did you really think you would have a point in your life where you just "arrived" or "made it?" We must continually and intentionally challenge ourselves to expand the boundaries of our world-view and not be conformed to the prison of familiarity.

Great leaders always stay in a place of learning. Leaders build an environment conducive to learning with great people, the best books, videos, schooling, classes, coaching, and courses they can take to help sharpen their leadership skills. There is never a point where we can rest from our learning? If so, our skills will atrophy and our people will suffer.

Now, Legson loved reading and devoured every book that he was given. He was given an English Bible and a copy of John Bunyan's *The Pilgrim's Progress.* However, he particularly loved Booker T Washington and Abraham Lincoln. He wanted to bring the liberation of Lincoln to his people oppressed by the evils of colonialism. Legson believed that the path to providing this liberation was through education.

He decided to become a teacher and free the minds of his people as he had experienced. "I saw the land of Abraham Lincoln as the place where one literally went to get the freedom and independence that one thought and knew was due him. One day I would also go there, I would also go to school there, and I would also return home to do my share in the fight against colonialism."—Legson.

Legson decided to go to America to study after learning that only a handful of the 3 million people in Nyasaland graduated from an American college. "I will be the first. I will be the first." He told himself continuously despite the obvious seemingly insurmountable obstacles in front of him.

So, what does a 16-year-old young man with no money, no transportation, no scholarship, and no idea of where America is, do? He starts to walk, of course.

On October 14th, 1958, Legson said his goodbyes to his village, and with some food, an ax, and his school uniform embroidered with "I Will Try," he departed for America. His mother laughed at him when he left and later admitted that she believed he would be gone for a week before he came to his senses.

Most people won't even consider taking the first step until they have everything they need. Nine thousand four hundred miles from his home in Washington State, and with enough food for five days, an ax, a world map, and a book *Pilgrims Progress from This World*, he began fulfilling his decision by walking.

> From October 1958 to January 1960, he made it to Uganda after walking 800 miles, working small jobs along the way to feed himself and earn enough money to get a ticket. Thus, I

arrived in Uganda, my temporary destination, but America, my goal, was still far away across land and ocean.

I pulled out my map when I was beyond the sight of the suspecting customs agents. With my fingers, I compared the distance I had covered and the distance still to be covered. I had come that far, and there was no point in returning now; I tried to console myself.

At the same time, I pitied myself for having plunged into such a journey. The motto on my shirt still said, "I Will Try," and I repeated the words as I had done now times without number: I Will Try. My shirt, however, was dirty, my shorts were dirty, and I was dirty.

While in Uganda, Legson didn't know what to do next, so he started doing odd jobs to prepare to continue his journey to Egypt. One day he found a library to find out any more information about studying abroad in the United States. He came across a directory of American universities, and the first entry he saw was Skagit Valley College in Washington State. He decided on Skagit and submitted an application letter.

Hoping for the best, he went back to work, and weeks later, amazingly, he received a reply from Skagit. His application was approved, and he also was awarded a full scholarship. His decision, fueled by continually affirming self-talk, was beginning to create the reality he desired. He still had to overcome the obstacle of getting to Washington State, but the miracle was already in motion.

His first step in getting to America was to attain a Visa. Legson would have to travel to Sudan to attain the Visa. The first portion of his trip was 800 miles, but this next journey would be 1600 miles.

Invigorated by the news of the scholarship, he continued his journey on foot toward Sudan.

He faced lions, hyenas, snakes, and elephants on his path and overcame every challenge. By the time he reached Sudan, two years had transpired. He crossed four countries and learned over ten languages.

Once he reached the US Embassy, he expected to face more obstacles to attaining his Visa. On the contrary, they were so impressed by his journey that they arranged all the documents he needed to travel and coordinated with Skagit College to get him on the next plane to America.

Over 2,500 miles on foot, indescribable pain and adversity in the hope of an impossible dream. On December 20th, 1960, that impossible dream became a reality when he was greeted at Seattle-Tacoma Airport by his host family and numerous school officials from Skagit College.

Legson Kayira would go on to graduate and become a best-selling author dedicating the rest of his life to liberating and educating his people. "They would reach their destinations sooner and merely by sitting down. I would reach my destination later and merely by counting my steps, but someday I would sit down and console myself that we both had reached our destinations, and this was all that mattered."

Reflecting on all that Legson has accomplished and the environment he set out to do the impossible, why did it work for him? What was the determining factor? What environmental and circumstantial impossibilities faced him? How did he overcome them? When we keep determined and focused on a goal, no matter what life throws at us, we have a chance! Legson had every reason to give up.

He had every reason to look at the hand he was dealt and choose to fold. Legson decided to play.

Legson won the fight because he decided the fight was won before he took one step. He didn't consider the possibility of failure. He resolved within himself that he would start to walk and the only acceptable destination would be in America at a university.

Our words, our declarations, combined with the certainty of a decision, render the myth of impossible, powerless. Excuses are the language of the mediocre. Being a leader means being resilient and dedicated and possessing a level of grit that will persevere in any situation. The way we win the fight over the limitations of our environment is to make a decision that we will not be defeated despite the circumstances.

As a leader, you will be put in environments that are not optimal for success. You will be placed in situations that seem impossible. You will find yourself in scenarios where lesser people will run away. Leaders decide to run to adversity, run into harm's way, and run to the hard because hard is authorized. That's why you are the leader. No matter the environment, the leader fortifies themself with a final decision that success will be attained. Decide to win the fight.

ROUND 6

THE CHALLENGER—SELF-IMAGE

He who learns must suffer. . .

—AESCHYLUS

We can live in our past, present, and future dimensions. Where we live primarily is the result of our inner voice. We are constantly trapped within two dimensions of existence; overwhelmed by our future and underwhelmed by our past. The majority of our negative self-talk resides in these two dimensions of existence.

Our past reminds us of our failures, mistakes, miscalculations, and bad decisions. Simultaneously, our future is painting a dystopian existence concerning events that haven't even occurred.

"I'm going to bomb on stage."
"I'm not going to do well on my presentation."
"People are going to think I suck."
"I'm not good enough to_____."

These two dimensions are pushing and pulling on our keys, getting them out of tune, and producing a sound that resonates with failure. Failure was imminent. I decided to go against the advice of my family after graduating high school with honors and multiple scholarships for the military. My parents didn't even want to sign the release form because I was still seventeen years old.

It was a difficult decision to turn down what they considered a sure thing for the uncertainty that comes with the military. I decided to join the Navy because I believed they would provide me with the right level of growth that I needed to become who I wanted to be.

My high school years had caused me to become irrationally arrogant at the prospect of my future success. I was near the top of my class, involved in nearly every extracurricular activity and multiple sporting teams. I scored very high on my SATs, allowing me to pick where I wanted to go to school.

As I decided on which schools to attend, my uncles would say, "smell myself." I recognized how severe my condition had become when I decided to walk out of the front door of my high school and drive to get lunch. This was not an approved activity for my school. No one questioned me at all because of who I was.

I was instantly terrified at the thought that no one was willing to hold me accountable for apparent injustices. I began to question the sincerity of my school leaders and inherently began to examine all of my choices. Everyone told me the right decision was to go to college, but I had never considered a different perspective.

A Navy recruiter briefed our class the following week and provided a different perspective. If everyone went to college directly after high school, what would separate me from them when seeking a career? What if I received the experience first and then acquired my degree. I thought this would give me a leg up on my *competition*, and I would grow up a little in the process.

I decided to join the submarine force because I thought submarines were incredible, and I couldn't swim, so if the boat went down, I wouldn't have to worry about being eaten by sharks. Not the soundest of reasons to pursue something so dangerous, but it worked for me.

I graduated boot camp months later and was immediately transported to Groton, Connecticut, just as winter began. I never knew how loud the voices inside your head could become until I got to Groton—cold, alone and uncertain.

I hadn't questioned my decision to join the military until this point. For me, Groton may as well be a foreign country. Nothing was familiar to me; they didn't even have a *Church's Chicken*. And again, the weather was just so cold.

One night, sitting in my room alone, the voices of defeat and regret began to override the whisper of triumph. I began to regret my decision to join the military.

My grades in school began to deteriorate, and my behavior in class began to be divisive and disruptive. I would talk to friends from home and in school. I would get so angry. I thought I was missing out on the fun they were having because I decided to join the Military.

As I slowly began to believe that I made a mistake being here, my belief system began sending messages to my consciousness, and as a result, I began to sabotage myself and fail. For the first time in my life, I failed an exam, then two. I distinctly remember the most prevalent phrase in my head during this time. I don't belong here.

I was up for an academic review board, and my instructors were baffled concerning me. I was put on mandatory study time and given a written reprimand. I had gone from a cumulative average of 96 to failing two consecutive exams.

I was trapped in an endless cycle; my belief formed the words *I don't belong here,* my mind repeated the phrase repeatedly. Ultimately, my actions confirmed those words. That confirmation came just a few days later.

My friends took me to a club. They thought alcohol would help me overcome my frustrations and disappointment—the perfect answer to an already bad situation. Let's add alcohol to your depression, right? Isn't that the correct equation? Let's see what the outcome will be; it couldn't possibly get any worse than it is now, right?

As expected, it had the opposite effect of what was intended. The voices only became louder and more prominent with every drink I consumed. Stumbling back to my room, I was convinced I was going home by any means necessary.

The opportunity to go home would make its way known quicker than I expected. After leaving the bar, we got into an altercation with several other people—culminating in a fight. We were all arrested and charged with assault.

Spiraling down a never-ending abyss, I saw no light. I had never felt darkness before. There was a genuine and palpable crushing that made it difficult to breathe. I could feel my heartbeat through each breath.

I sat in that jail cell, hoping that the rush of tears would somehow drown me so that I didn't have to deal with this anymore. I was so used to having the answer in every situation, but I couldn't even get to an emotional state that would allow me to consider a way out.

My mind would travel to my presumed future, filled with disappointed looks and comments from my friends and family. Every look and comment was a dagger that I felt in the present. Out of the darkness, the words I heard changed from "you don't belong here" to "I told you that you don't belong here." The words that I consistently spoke to myself led me to the only possible conclusion.

Once I got out of jail, tail between my legs, I called my dad to tell him what had occurred and that I would eventually return home—a failure.

He patiently listened while I recapped the past few months through episodes of screaming, crying, yelling, sobbing, and finally, defeat. He just asked if I was done, and what he said next never left me:

> Is this who you have become? The son who left here was obsessed for months with winning a dance contest because I told him he couldn't dance; where did he go?

> The son who left here would spend hours a day working on his free throws because he missed his shots during a field day—only to come back the next time to make all five shots and win the trophy for his team. Where did he go?

> The son who left here did not know how to quit.

> I guess I was right; you should've gone to college. You come home and prove to everyone in your family that you thought you made a mistake—that they were right. Is that what the Navy taught you—to quit? Is that what you want?

My dad had a way of getting under my skin like no other. When I was in 5th grade, I was completely obsessed with Michael Jackson and MC Hammer. I was completely convinced that I could dance just like them. One day I decided to present my phenomenal dancing skills to my parents.

I perfectly merged the footwork of MJ with the energy of MC Hammer. It was a thing of beauty in my mind. I turned on the cassette player (for those of you under the age of 35, this was the mechanism we used to play music in the 80s and 90s) and went into my dance.

Afterward, assured that I had wowed my parents, my father looked directly at me, shrugged his shoulders, and told me he thought it was mediocre. I was dejected and defeated, but something ignited and caused me to practice even harder.

There was a dance contest for our final week of school, about six weeks away. I locked myself into my room for the entire time, practicing every move with absolute precision. I borrowed the VHS tape (a VHS tape was played on a VCR which was how we viewed movies and videos in the 80s and 90s) from my aunt to practice with. I practiced so much that I broke the tape and the player.

I didn't say anything to my father for the six weeks outside of general pleasantries and acknowledgments. When the day of the dance contest finally came, I was ready. I went to the cafeteria, turned on the dance floor, and did my thing. I won the dance contest and immediately left the school running full speed to get home and show my father.

I ran into the room and yelled, "You thought I couldn't dance!" while throwing the trophy on the bed. He glared at me with this evil grin while I was out of breath and feeling rather accomplished.

He said, "I knew you could dance, your mother told me about the dance contest, and I wanted to see how good you could get."

Confused, I managed to mutter an incoherent "huh." He stood up and told me, "I know you better than you know yourself. I know what will get the best out of you. You refuse to let anyone tell you what you can and can't do. I just pushed you a little."

Continually throughout my life, my dad would strategically trigger me, just like this dance contest, to ignite my competitive nature. He knew exactly what to say to get me back on track. Regardless of how often he would pull this card, I would fall for it.

Now, flash back to my room at Groton and me responding to my father. "Is this what I want? Is this what I want? Are you kidding me?" Furious, I hung the phone up on him and smashed it on the floor of my room. I yelled at the broken phone, "THIS IS NOT WHO I AM!!"

Amazingly enough, the negative voices stopped. Everything just stopped. It was the first time I had silence in months. I finally clarified my situation because my dad reminded me of who I was. Getting a clear picture of who I was allowed me to silence the negative voices and turn up the affirming ones.

As a leader, you must be willing to fight for your identity. You must be willing to look in the mirror and aggressively reaffirm to yourself who you are.

One of my favorite movies is Black Panther. Self-admittedly I am a huge geek for all things Marvel. You can learn a lot about leadership from a comic book. A powerful lesson illustrating the truth of the power of affirming your self-image was revealed in a scene within the movie.

T'Challa was the Black Panther who received the mantle from his father, that was the Black Panther before him. While at the world council to discuss the collaboration of the newly revealed superheroes and the world governments, a terrorist explosion killed his father. This required the mantle of leadership of Wakanda to be passed to T'Challa.

To assume the mantle of leadership, T'Challa had to have his supernatural powers stripped and fight any challengers from each of the tribes within Wakanda. To lead Wakanda, you had to prove that you could be the protector and king. During the ceremonial gathering of all of the tribes, unexpectedly, M'Baku from an outlying tribe decided to challenge the throne.

T'Challa agreed, and they began to battle. M'Baku was considerably stronger and bigger than T'Challa. T'Challa started strong, but as the fight continued, M'Baku began to overwhelm T'Challa with his strength, making it look bleak for him. Worry and anxiety swept through the people as T'Challa lay on the ground in pain.

Cutting through the silence, the powerful voice of his mother, Ramonda, the Queen of Wakanda, shouts, "Show him who you are!" Hearing those words, you see T'Challa connect back to who he knew he was and gain the upper hand on M'Baku. Almost like gaining a supernatural injection of strength and resolve, he outmaneuvered M'Baku and utilized his strengths to defeat M'Baku and become King of Wakanda ultimately.

On this leadership journey, we will face battles that seem overwhelming and impossible and will challenge our self-image. The challenges will cause you to question who you are. Are you good enough? Are you strong enough? We must remember who we are. We must recall the victories we have had. We must consistently reaffirm what we stand for and who we stand for.

Now, back to my predicament. Was I still in trouble with my chain of command from that night? Yes. Were those failed exams still looming? Yes. What changed was the story I told myself about those things. I told myself that these were not indications of who I was but bumps in the road from which I was meant to learn—and grow.

There is a continuous narrative running in the background of our consciousness. The tape is endless. For so much of our lives, someone else has controlled the tape.

The narrative of our lives tries to be written by someone else. It is easy to forget that we control the soundtrack. It's easy to forget that we have the pen to construct our own narrative. There may be chapters in the story that don't necessarily work out in the hero's life, but the chapter isn't the book. We may all be amid a bad chapter, but we can't allow that to define the entire story.

Remember the challenger is our self-image and what we focus on and listen to will direct our path. In leadership, you are charged with helping others become the best version of themselves. To accomplish that mission we first must be the best version of ourselves.

What do you what your self-image to say about yourself? What are you going to follow as a leader? In the end, what you want is meant to be worked and trained for, not merely hoped for. The challenges never end, but your ability to look inwardly and say to yourself, "self, this *is* or *isn't* who you are." That saying is your fight to lead.

ROUND 7

THE CHALLENGER—PERSPECTIVE

Raise your words, not your voice. It is rain that grows flowers, not thunder

—RUMI

O ver the ages, the importance of what we say to ourselves has been held in the highest regard. In ancient Greek philosophy, Heraclitus (540-480 B.C) defined the power of our words to be the "universal principle through which all things were interrelated, and all-natural events occurred."

We use our words or language to represent our experience of the world, and the words we use encompass us and shape the world in which we exist. According to a Jewish philosopher during the time of Jesus, Philo described the power of our words to be the "intermediate between ultimate reality and the sensible world."

Our experience of the world is not the world in the same way that the map isn't the actual territory; it is merely our representation or perception of the world at that moment. Our words provide depth, color, and contrast to our life experiences.

As leaders, we must maintain perspective and clarity about all situations. With so many possible scenarios we have to contend with daily, it becomes easy and comfortable to believe everything we think. We can convince ourselves of anything, especially when it removes us from difficulty into a benefit.

As a Chief Petty Officer, I am responsible for young and growing young men and women trying to navigate their lives. Because they are young and growing, they often do very childish things, and we, as leaders, are responsible for their performance.

One instance that stands out to me is when one of my Sailors was arrested for drinking and driving. At 3:34 am, I received a call that my Sailor was in holding and someone needed to come and pick him up. I am, at this moment, extremely frustrated and angry because this is an area we emphasize to our Sailors the responsible use of alcohol.

Not only am I mad at the Sailor, I know that my leadership for my *failure* will fiercely judge me. In the military, it is often the policy to hold the supervisors accountable for the decisions of the people they lead. This is the chain of command structure, good, bad, or indifferent.

So two realities are converging upon me while I am in my car at 4:10 am to pick up my Sailor. My career is now in jeopardy, which is an exaggeration generated by anxiety and fear. Two, the overwhelming number of times I said, "I can't believe the Sailor would do this to me."

I said those words repeatedly, and it only fueled my anger and drove me to only think about how this situation would affect me. I allowed the words that played in my head to detach me from any clarity or perspective about the situation. I wasn't considering whether or not the Sailor was ok. I wasn't feeling the circumstances surrounding the arrest; was it problematic or unwarranted?

If it weren't for my wife, who was driving with me, I would have allowed the reality my emotions had constructed for me to dictate my behavior. It wasn't until she looked at me and said, "Well, he didn't wake up this morning and say I want to get a DUI today." I immediately snapped out of my ego-filled rage and began seeing things with empathy.

Left unchecked, I would have considered this vicious act as aimed towards me—as if he was targeting me directly just to put my career in jeopardy. That is such a *victim mindset* it stopped me dead in my tracks.

This is just a minor example of how our words can shape our experience of the world, and we do a massive disservice to the people we lead when our words negatively compromise our perspective and clarity.

The inventory of the words we carry directly reflects our positive or negative experiences. Two people experiencing the same occurrence in the same place and at the same time can have two very different perspectives on what actually occurred.

Every day we communicate with ourselves and others, but we are typically unaware of the arrangement of the words we use to represent our experiences. What is the difference between "The bee stung Julie" and "Julie stung the bee?" The words are the same, but how they are arranged is different.

The meaning of the experience is determined by the arrangement of the signals provided to our brains. Our brains are supercomputers, and our words, our inner voice, give the commands for our programs to run. If we program the commands in the proper order, our brains will use all of their resources to produce the result we desire. The Reticular Activating System controls this process in our brain.

What is the Reticular Activating System or RAS system? A bundle of nerves at our brainstem filters out unnecessary information so only the important stuff gets through. It starts above your spinal cord, about two inches long. It's about the width of a pen and where all your senses (except smell) are. It is our own personal life spotlight. It is illuminating the things necessary to us and filtering out the things that aren't.

RAS is responsible for the phenomenon of noticing your new car, your new dress, and your new shoes *everywhere* after you have purchased them. Based on your desire to acquire these items, your brain, through the RAS, prioritizes your awareness to notice or highlight them whenever they come within your purview.

All of this happens below our awareness. The RAS program itself works in your favor without you actively doing anything. Our RAS is impartial and runs continuously. The RAS responds to our most emotional, definitive, and consistent messaging. It can work for us or against us; the choice is ours.

In the same way, the RAS seeks information that validates our beliefs. It filters the world through the parameters we give, and your beliefs shape them. If you believe you are a good writer, you most likely are. If you think you are bad at math, you probably will be.

The RAS helps you see what you want to see and, in doing so, influences your actions. This is so important as leaders because, through RAS, we can literally direct our brains to focus on what we want to obtain.

This was brought to light for me during a relatively normal occurrence. I had just flown into LAX for a conference. This was my first time coming to LA, so I was unfamiliar with the number of people and traffic at the airport.

I pull up my Uber app and schedule a ride to the airport. Once I made my way from picking up my luggage to the rideshare pickup area, I was slightly overwhelmed by the volume of people and cars. The car that was picking me up was a black Nissan Maxima with license plates starting with WXT.

As I found a place to stand and wait, I kept saying to myself black maxima. Over the next few minutes, as I was waiting for the car, every black or Nissan car came to my present awareness. It felt like the scene in *The Matrix* where Neo and Morpheus were calling up anything they wanted into reality.

RAS allowed me to filter out all of the cars that didn't fit into my description and only focus on the car I needed. I even took it a step further to filter it down to the license plate, and all I began to notice were black Nissan with license plates starting with W. I was sure that this phenomenon had happened to me before, but this was the first time I intentionally used it on my own behalf. I wanted to control and filter what I wanted to see when I wanted to see it was a fascinating thought as I stepped into the black Nissan Uber.

Understanding how our minds work and process information is vital as a leader because our perspective dictates our behavior. The correct assessment of our circumstances combined with the words we utilize to describe and represent that occurrence to ourselves is the key to success or failure as we lead.

Our words are programming languages for everything that we do. The command we send our brain, "I always lose," sends our brain to search for confirming evidence to support that command.

If I send the following command to my brain, "I always lose." What are the chances that my experience will be positive?

Our brain then opens our catalog of mistakes, failures, bad decisions, and losses. The brain transmits that information to our consciousness to close the loop on the command. When our brains transmit the files to our consciousness, it also sends the associated emotion. Based on how we interpret the information, the emotion is then classified into a feeling. The most likely feeling associated with losing will be negative in this case.

So not only do we receive the evidence of losing, but we also receive the feeling associated with losing. We experience the moment's anger, resentment, and frustration, which sets our emotional state.

Growing up, I played many sports, but baseball was my favorite place where I had the most talent. When I reached high school, I was one of the top players in our district. I hoped to go to the major leagues one day, but being 5'10" and 137 lbs wasn't necessarily the best metric to reach that goal.

During my playing days, I distinctly remember that how I started a game would directly correlate with how well I played the entire game. If I struck out during my first at-bat, that would dictate my emotional state during the rest of the game, and I would always play below my potential.

I would get down on myself and say that *I let down the team, that I'm not as good as I think, I'm horrible,* etc. These words would carry over into my performance in every area of the game, and inevitably, I'd strike out again or commit an error because I wasn't the best version of myself.

Instead of utilizing RAS to extract the good out of the situation, like recognizing even though I struck out, the pitcher tipped off his fastball, which let me know what to expect next time I came up, I would only extract the bad so that is precisely what I would experience.

What happens when I send the command, "I always win." Our brain opens the catalog of wins, transmits that information to our consciousness, and we become aware of our victories and begin recognizing opportunities to win. This action alone widens our aperture and makes everything around us a resource rather than an instrument of doom.

One time when this saved me as a leader was when some vital equipment malfunctioned right before our submarine was meant to get underway. Typically, before the sub gets underway, everyone verifies that all systems are operating correctly and aligned for operations. This time I was away from the submarine taking care of a personal matter, so I trusted my team to ensure everything was operational and aligned.

As I was getting back to the submarine with a little less than an hour left before we were to go, my entire team rushed to me in a state of panic. "Chief! Chief, the control panel is not getting any power, and all the fuses are blowing!"

My immediate desired response was absolute dread and terror. I overcame that initial wave of emotion by nodding my head in agreement and saying, "Interesting." At that moment, it was important to disarm the power of the emotions and thoughts that drove me to begin to speak disempowering language to myself. I used the word interesting because it isn't good or bad; it's just…interesting. This left me the space to define the situation how I wanted to.

Oftentimes, when your team brings you bad news, they are inadvertently transferring their negative emotional state from them to you. I call this D.A.T.S. *Don't absorb the stupid.* Unchecked, you become a mirror for their emotions, and you are emotionally compromised and lose the perspective needed to resolve the problem at hand.

As my team is in front of me yelling and panicking, I calmly ask them, "what are our options? This is an easy problem to fix." It, in fact, was not an easy problem to fix, but they didn't know that. I didn't want them to start in a place of defeat before they even tried, so I told them that this was an easy problem to resolve, and their moods changed almost immediately.

They were just in the midst of panic, and as the leader, I gave them permission to put their fear aside and approach the solution reasonably. Once they calmed down, I told them, "there hasn't been a problem we couldn't resolve, and we aren't starting now." Just as I went to walk away to go back down to the submarine, my youngest Sailor blurted out, "what if we were using the wrong fuses?"

I turned back to look at him. He continued, "Even though we are pulling fuses out of the right container, what if there was a mix-up? What if the fuses in the container had been placed in the wrong container, and that's why it isn't powering up."

I stopped in my tracks and looked at my team. They all ran past me and down to the panel. After conducting all of the verifications, he was right, and we could get underway without issues. As a result of the team seeing with clarity and no longer in a panicked state, they could get to the solution quickly.

Readers, if you are a leader, The leader must be cognizant of their team's emotional state before absorbing that. Don't become part of the problem. We must be careful of our words so that we are directing them toward success rather than failure.

One of the most illustrative examples of the power of one's words in creating reality is shown through the life of Nelson Mandela. Nelson Mandela was born in 1918 in a small village in South Africa, and he was a member of the Xhosa tribe and grew up in a rural area where traditional customs and values were still strongly upheld.

Mandela attended a local mission school and later studied at the University of Fort Hare, one of the few institutions of higher education for Black students in South Africa at the time. After completing his studies, Mandela worked as a lawyer and became actively involved in the anti-apartheid movement. Apartheid was a vicious and evil system of racial segregation and discrimination that was implemented in South Africa from 1948 to 1994.

The word "apartheid" is Afrikaans for *separateness* or *apartness*. Under apartheid, South Africa was divided into different racial groups, with white people at the top, followed by people of mixed race, then Black people, and finally, Asian and indigenous people at the bottom.

Each racial group had a separate legal status and was restricted from living, working, and traveling in certain areas. Apartheid devastated the lives of Black South Africans and other marginalized groups in South Africa. The racial segregation and discrimination system was morally reprehensible and had a wide range of adverse effects on individuals and communities.

Apartheid created deep social, economic, and political inequality; it was a system that dehumanized and oppressed most of the population. Black South Africans were excluded from many types of employment and were often paid less than white workers for the same jobs.

Schools for Black children were underfunded and overcrowded, and the quality of education was poor. This led to high dropout rates and low levels of literacy, which in turn contributed to the economic disadvantage they faced. They were often denied access to adequate healthcare and were subjected to discrimination by healthcare providers. This led to high rates of preventable diseases and poor health outcomes.

In the 1940s and 1950s, Mandela was a leader in the African National Congress (ANC). This organization advocated for the rights of Black South Africans and sought to end the system of racial segregation known as apartheid. In 1952, he opened the first Black law firm in South Africa with his partner Oliver Tambo. Mandela knew that it was his life's purpose to eradicate Apartheid.

In the early 1960s, the South African government banned the ANC, and Mandela was arrested and charged with sabotage and conspiracy to overthrow the government. He was found guilty and sentenced to life in prison in 1964, and he was sent to Robben Island prison, where he spent the next twenty-seven years. While in prison, Mandela became a symbol of the anti-apartheid movement, and his name became synonymous with the struggle for freedom and equality in South Africa.

Nelson Mandela was 46 years old, a point where most would be thinking about how to approach the end of their work towards retirement. But this was just the beginning for Mandela. He was the living embodiment of "to whom much is given, much is required."

In the fight for what was right, in the battle for equality, he was imprisoned for what was believed to be a life sentence. In the eyes of those he led, all was lost, and there was no reason for Mandela to believe anything different. Robben Island was meant to be where he came to die, along with his dream of eradicating apartheid. The government meant to bury him and anyone who shared the desire for freedom. What they didn't realize was that they were seeds. In the darkness of prison, Mandela drew from every resource in his reach to fuel his dream.

The power of perspective is that it is solely the individual's responsibility. Nothing external is required; only choice. When we become overwhelmed by a situation, it is easy to forget that we retain the ultimate power to choose how we will respond.

Author and speaker Wayne Dyer said, "We see things not as they are, but as we are." Before becoming the first black president of South Africa, he said he had a dream of leading his country one day. Knowing this dream to be true, Mandela used his time in prison to change himself into someone who could lead his country. He said, "one of the most difficult things is not to change society but to change yourself."

Robben Island was the place that transformed Mandela. He had years of self-examination and meditation— seeing positive things in his darkest hours. He didn't stop trying because he knew the truth and stayed with that reality. Without the darkness of prison days, Mandela might never have become such a remarkable leader after he walked free. He had gone to prison as an angry rebel who believed that violent revolution was the only answer.

"At least, if for nothing else," he wrote in a 1975 letter to his wife, "the cell allows you to look daily into your entire conduct, overcome the bad and develop whatever is good in you. Mandela was able to take control of his reality during his prison sentence and form himself into the person he needed to be not only to survive but thrive.

The secret to his remarkable transformation was contained in a poem he repeated to himself every day, multiple times daily.

Invictus by William Henley

Out of the night that covers me,
Black as the pit from pole to pole,
I thank whatever gods may be
For my unconquerable soul.
In the fell clutch of circumstance
I have not winced nor cried aloud.
Under the bludgeonings of chance
My head is bloody, but unbowed.
Beyond this place of wrath and tears
Looms but the Horror of the shade,
And yet the menace of the years
Finds and shall find me unafraid.
It matters not how strait the gate,
How charged with punishments the scroll,
I am the master of my fate,
I am the captain of my soul.
'I am the master of my fate; I am the captain of my soul.'

Mandela knew that the person who went to prison was not the person who would survive. Through his intelligence, refined inner dialogue, and dignified defiance, Mandela eventually bent the most brutal prison officials to his will, assumed leadership over his jailed comrades, and became the master of his prison.

He emerged from it as the transcendent leader who would fight and win one of the greatest political battles of the century that would create a new democratic South Africa. Mandela's mastery of his communication with himself ultimately created the reality he desired.

We leaders must go boldly into the darkness and be the light. We are charged to gaze fearlessly into the chaos and command order. We are charged to extract the 0.001 percent from the 99.999. Nelson Mandela did not negotiate with the reality he was given, nor can you.

The fight of perspective is not an easy foe; it looks imposing, impossible, and immovable. The difficulty of victory is found in its simplicity; make the choice that defeat is not an option.

ROUND 8

THE CHALLENGER—NEGATIVE VOICES

*I cannot control everything that happens to me; I can
only control the way I respond to what happens. In my
response is my power.*

—ANTHONY ROBBINS

Behavioral scientists have recently determined that we have heard over 200,000 no's from birth to eighteen years old. Additionally, research has determined that approximately eighty percent of what we think is negative, counterproductive, and actively working against our desires. We are challenged with overcoming eighty percent of our programming being negative.

Everything and everyone around us programs us. Most of the time, it's the wrong type of programming. This programming that we receive and still receive comes to us unconsciously from our parents, siblings, teachers, schoolmates, friends, associates, media advertising, and the news.

Year after year, our life's scripting is being written without our concurrence. Our self-image was being developed and cultivated without our permission. As time progressed, we began to join in sculpting our image and writing our script. We believed the words we were being told and the words we said to ourselves—were true.

We began to hear the exact words and thoughts repeatedly. Thousands of times, we were told what we could not do, what we could not accomplish, and what we could not achieve.

Repetition is a convincing argument. At some point, we believed these words and lived out the reality of those words. We became what we most believed about ourselves:

"As a man thinketh, so is he," Prov. 23:7.

Earl Nightingale said, "We become what we think about most of the time, and that's the strangest secret."

Seneca said, "Everything hangs on one's thinking…A man is as unhappy as he has convinced himself he is."

We are now scientifically catching up to these principles written thousands of years ago. Our success or failure in any endeavor depends on our programming. Our brain believes the story we tell the most and creates that reality for us to live in. We now understand how to communicate with ourselves to eliminate negative self-talk and create the reality we desire.

Our negative self-talk is not a reflection of reality. Our negative self-talk condemns us and highlights our failures and inconsistencies. It can paralyze us and bring our progress to a screeching halt.

Research has shown that self-compassion is one of the most effective ways of reducing negative self-talk. Our normal inclination when fighting a battle is to fight and resist. We try to ignore the voice. We try to disavow the voice. But the voice persists and becomes louder and more negative. The truth is; that what we resist will persist.

We will continue to lose the battle if we fight through traditional means. Instead of resisting, we must first become aware of what the voice is saying and then acknowledge it.

When we acknowledge our feelings at the moment, we get the opportunity to determine what is true. We can give ourselves the benefit of the doubt. Just as you would validate a friend's feelings if they were hurt, do the same for yourself and acknowledge your pain or discomfort.

An essential part of self-compassion is not just replacing negative thoughts with positive ones but intentionally becoming aware of your emotional state. As you become aware of the voice, acknowledging how you feel in the moment, continue to have compassion for yourself by recognizing that you are not alone in your suffering.

We all suffer to varying degrees. This is a part of being human and understanding that we are merely human and prone to mistakes, but we ourselves are not mistakes. This is how we begin to have compassion for ourselves and practice a non-judgmental understanding of our own pain, suffering, inadequacies, or perceived failures.

Use this statement to speak to yourself lovingly, compassionately, and kindly. "I am imperfect because I am human. I am a work in progress. I am not a failure." This statement disarms the power the negative voice has over us. This statement reduces the judgment and shame we experience when we hear negative self-talk. Self-compassion renders negative self-talk powerless.

When we no longer judge ourselves for our shortcomings, the truth becomes clear:

"Failure is an event, not a person. I AM ENOUGH."

If you accept that we are a work in progress, then all of this takes effort and time, right? Well, what if I told you that it would also impact your health if you don't start? Specifically, what if your head impacts your heart? It's not just the negative self-talk that is bad for your mental health. It causes stress that, if not changed, will actually damage your heart.

"Petty Officer Cleery, if you don't make major changes now, your health will deteriorate dramatically to a place that we may not be able to recover from."

As John left the doctor's office, he reflected hard on the words of his physician and pictured his wife and two children. He thought about not being able to see his son play his first high school soccer game. He thought about not being able to harass the young man that would take his daughter to the prom. He thought about his upcoming 20th anniversary with his high school sweetheart. He knew he needed to make changes but didn't know where to start.

His performance at work began to diminish considerably about eight months ago. John was placed in charge of his division onboard the USS John Finn. He was a top performing, early promote (early promote is a designation for the top-ranked ten percent of all Sailors) Sailor on track to be promoted to become a Chief Petty Officer. The only thing he was missing was to lead his own division successfully.

Prior to leading his own division, PO Cleery was very comfortable with his own performance and knew how to get the best out of himself, but he had no idea how to transmit that same success to the people he was chosen to lead.

Like most new leaders, John attempted to force his process and system on his team of twenty-three people, and initially, the results looked promising. But as the stress of upcoming operations began to build, he shifted from the calm, deliberate, and instructive leader to an irrational, angry, and unreasonable dictator.

He struggled to bring his best to work, lacking clarity and connectedness with his team. His team feared his outbursts, and despite multiple sit-downs with leadership, nothing was getting better. His team's results dropped by forty-two percent, and he was given a letter of instruction, effectively placing him on probation.

Meanwhile, John wasn't getting any sleep at home, and his relationship with his wife was suffering because the same anger and frustration with work were being brought into the home. John was spiraling emotionally, mentally, and physically with no way to pull out of it.

Before he left the doctor's office, the physician gave him the contact information for Heartmath Institute and told him that he believed the source of all of his issues was stress related, and they might be able to help.

For more than twenty-five years, the *HeartMath Institute Research Center* has explored the physiological mechanisms by which the heart and brain communicate and how the heart's activity influences our perceptions, emotions, intuition, and health.

HeartMath Institute is one of the most innovative companies investigating one of the most detrimental epidemics in the world today, stress. A growing body of compelling scientific evidence demonstrates a link between mental and emotional attitudes, physiological health, and long-term well-being.

This is what the institute has to say:

- 60% to 80% of primary care doctor visits are related to stress, yet only 3% of patients receive stress management help.
- In a study of 5,716 middle-aged people, those with the highest self-regulation abilities were over 50 times more likely to live without chronic disease 15 years later than those with the lowest self-regulation scores. Positive emotions are a reliable predictor of better health, even for those without food or shelter. Negative emotions are a reliable predictor of worse health, even when basic needs like food, shelter, and safety are met.
- A Harvard Medical School Study of 1,623 heart attack survivors found that when subjects became angry during emotional conflicts, their risk of subsequent heart attacks was more than double that of those who remained calm.
- A review of 225 studies concluded that positive emotions promote and foster sociability and activity, altruism, strong bodies, and immune systems, effective conflict resolution skills, success, and thriving.
- A study of elderly nuns found that those who expressed the most positive emotions in early adulthood lived an average of 10 years longer.
- Men who complain of high anxiety are up to six times more likely than calmer men to suffer sudden cardiac death.
- In a groundbreaking study of 1,200 people at high risk of poor health, those who learned to alter unhealthy mental and emotional attitudes through self-regulation training were over four times more likely to be alive 13 years later than an equal-sized control group.

- A 20-year study of over 1,700 older men conducted by the Harvard School of Public Health found that worry about social conditions, health, and personal finances all significantly increased the risk of coronary heart disease.

- Over one-half of heart disease cases are not explained by the standard risk factors such as high cholesterol, smoking, or a sedentary lifestyle.

- An international study of 2,829 people ages 55 to 85 found that individuals who reported the highest levels of personal mastery – feelings of control over life events—had a nearly 60% lower risk than those who felt relatively helpless in the face of life's challenges.

- According to a Mayo Clinic study of individuals with heart disease, psychological stress was the strongest predictor of future cardiac events such as cardiac death, cardiac arrest, and heart attacks.

- Ten 10-year studies concluded that emotional stress was more predictive of death from cancer and cardiovascular disease than from smoking; people unable to manage their stress effectively had a 40% higher death rate than non-stressed individuals.

- A study of heart attack survivors showed that patients' emotional states and relationships in the period after myocardial infarction were as important as the disease severity in determining their prognosis.

- Separate studies showed that the risk of developing heart disease is significantly increased for people who impulsively vent their anger and those who tend to repress angry feelings.

We are experiencing an epidemic of epic proportions. We recognize that stress is an immense fight, but the problem is that primarily in leadership, we are fighting the wrong opponent.

Typically, in work settings, personnel are evaluated based on their behaviors. Interestingly enough, to correct those behaviors, we communicate that they should fix their behavior. During individual counseling, strengths and weaknesses are highlighted, and we tell people to do more of their strengths and less of their weaknesses.

One of the wisest men in history, Albert Einstein, said "we couldn't solve our problems with the same thinking we used to create them. Simply telling people to correct their behaviors through altering their behaviors has proven unfruitful."

As humans, we have spent our lives operating a certain way, and these patterns are ingrained within us and are hard to break. Many times we attempt to overcome our limitations through sheer willpower. This strategy becomes unsustainable as our energy wanes, and we further contribute to our stress levels with diminished energy.

How do we win this fight?

Over the course of a few months, HeartMath helped Jon to see that the root of his decline in performance was due to a deterioration in the management of his physiology. When we are in a constant state of stress, we operate primarily out of survival mode.

Survival mode, also known as freeze, flight, or fight mode, is the state our nervous system shifts to when we perceive danger. This state was meant to protect us from lions, tigers, and bears, but our nervous system doesn't know the difference between an overdue project or a physical threat.

Research is overwhelming that stress is an epidemic wreaking havoc on the lives of leaders seeking to add value to their teams. Through his consultation, John was shown several tools to help him reduce his stress, calm his negative emotions, regain his clarity, and ultimately be the best version of himself by leveraging his physiology.

Leveraging our physiology is how we enlist our entire body to direct positive energy into our consciousness to overcome the negative results of stress. New research into understanding our physical, mental, and emotional systems has revealed some amazing information.

There is abundant evidence that negative emotions alter the activity of the body's physiological systems. The more we experience what we define as *negative*, our physiology becomes more out of sync and out of balance. The longer our physiology remains in this out-of-balance state, the more our feelings of negativity perpetuate.

This process becomes a cycle of destruction that wreaks havoc on our physical, mental, and emotional health. HeartMath Institute calls this state *incoherent*—to fight and defeat the stress and discomfort leading to subpar performance requires us to shift from our heads to our hearts.

At the center of regulating our emotions, feelings, and overall livelihood are, of no surprise, our hearts. Our heart is more than just the pump in our chest that sustains life. It has been revealed that it is an access point to a source of wisdom and intelligence that we can call upon to live our lives with more balance, greater personal effectiveness, and experience greater fulfillment.

Our heart is an information processing center with its functional brain that communicates with and influences the brain through the nervous system, hormonal system, and other pathways. As more research was conducted, HeartMath discovered that the rhythmic pattern of heart activity was directly associated with the subjective activation of distinct emotional states. They found that the heart rhythm pattern also reflected changes in emotional states, which covaried with emotions in real time.

They found strong differences between quite distinct rhythmic beating patterns readily apparent in the heart rhythm trace and directly matched the subjective experience of different emotions. In short, they found that the pattern of the heart's activity was a valid physiological indicator of emotional experience and that this indicator was reliable when repeated at different times and in different populations.

This observable pattern within the natural fluctuations in heart rate is called heart rate variability or HRV. Initially, HRV was used during pregnancies to determine whether or not the fetus was in distress. Further expansion of the research began to show that HRV was a valid indicator of emotional stress in adults as well.

Also, it was shown that raising HRV correlated to lower stress levels, more perceived positive emotional states, greater clarity, and higher self-regulation. For the first time, we can quantify and manipulate an individual's mental resilience and toughness level and not rely on subjective theories that are hit or miss.

HearthMath developed a process that would alter an individual's HRV and, through a simple exercise, use the heart fluctuations to bring them out of survival mode into a more optimal physiological state. Through a heart-based breathing process, the negative feelings and emotions associated with the elevated stress would diminish significantly in only 5-15 minutes a day.

Raising HRV through heart-based breathing has been proven to reduce stress, increase cognitive processing, increase overall well-being, and increase our self-regulatory system. This means our capacity to *bounce back* from low emotional points will increase dramatically.

Heart-based breathing is completed in four steps.

1) Place your hand on your heart and become aware of your heartbeat. You can also access pulse points to feel your pulse if feeling your heartbeat is problematic.
2) Close your eyes and bring to your mind one thing you are grateful for.
3) The goal is to inhale and exhale at the same rate. With that image in mind, begin to inhale for 4-7 seconds and exhale for 4-7 seconds. The optimal rate is 6 seconds inhale and 6 seconds exhale.
4) While breathing, concentrate and breathe into the area around your heart. 3-5 minutes is all it takes.

Heart-based breathing only takes a few minutes daily, but the benefits last a lifetime. Alone each method is powerful; combined. They are unstoppable.

PO Cleery gave these methods a try and, within the first couple of weeks, began to see improvement in his mood, clarity, and emotional regulation.

After a few months, he could regain his team's trust and increase his effectiveness and performance. He introduced this practice to his team, and they have collectively lowered their stress levels and grown closer as a team. Thousands of people and teams have utilized these simple techniques to fantastic success.

The enemy is our physiology, and the fight is to regain control of it and maintain balance. Any inconsistency within our physiology will communicate that dismay to our emotions and feelings. This dismay is then transmitted to our thinking and, ultimately, our behaviors and performance. It's the leader's responsibility to get to the root of the matter, and the root is leveraging our physiology for success.

As the conductor of the symphony, that is, our physiology, once we align the percussion session (our hearts), every other section receives their marching orders and plays accordingly. The hardest person we will ever lead is ourselves, and leading ourselves effectively means taking responsibility for ourselves and fixing what lies within.

Dedicating minutes per day to eradicating and replacing toxic programming could be the difference between leading a life of value or one that is nothing more than a liability to ourselves and the people we have the privilege to lead.

ROUND 9

THE CHALLENGER—HOPELESSNESS

*In some ways, suffering ceases to be suffering at the moment it
finds a meaning, such as the meaning of a sacrifice*

—VICTOR FRANKL

Victor Frankl said, "I did not know whether my wife was alive,
and I had no means of finding out (during all my prison life,
there was no outgoing or incoming mail), but at that moment,
it ceased to matter. There was no need for me to know; nothing
could touch the strength of my love, my thoughts, and the image of
my beloved."

Between stimulus and response, there is a space of time. We can
choose between something occurring in our lives and our immediate response. That space of time allows for a significant choice. The
stimulus is irrelevant; we can do nothing about it. What we *do* control, however, is what we choose to do next.

If we don't choose correctly, the innate emotion of the moment
will capture our perspective, and we will only see the situation
through that emotion. Leaders are charged with solving the most
complex problems, which elicit potent emotions, anger, fear, frustration, etc.

If we allow ourselves to become slaves to our emotions, we lose the ability to dictate the most favorable outcome. Our emotions mislead us. When this happens, we lose control of ourselves, and we lose control of ourselves, and we lose control of everything.

In 1939, Germany invaded Poland, and what followed was one of the greatest atrocities in the history of humanity; the holocaust. The Holocaust was the genocide of European Jews during World War II. Between 1941 and 1945, Nazi Germany and its collaborators systematically murdered around six million Jews across German-occupied Europe, nearly two-thirds of Europe's Jewish population.

One of the most prominent survivors of the Holocaust was the neurologist, psychologist, and philosopher Victor Frankl. Viktor Emil Frankl was born in Vienna, Austria, on March 26, 1905. He received his MD and Ph.D. degrees from the University of Vienna, where he studied psychiatry and neurology, focusing on the areas of suicide and depression.

As a medical student in the late 1920s, he successfully counseled high school students to eliminate suicide virtually. Because of these accomplishments, he was asked to head the suicide prevention department of the General Hospital in Vienna.

After treating thousands of people over the four years he was there, Frankl took a position as the head of the neurological department at the Rothschild Hospital, one of the few facilities that allowed Jews to practice medicine at that time.

In 1942, Frankl, his parents, wife, and brother were arrested and sent to the Theresienstadt concentration camp. Frankl chronicles his experience within the concentration camps in his book *Man's Search For Meaning*.

In the book, he describes in great detail the prisoners who gave up on life. They had lost all hope for a future and were inevitably the first to die. He states that they died less from a lack of food or medicine than from a lack of hope or something to live for.

Frankl states, "I kept myself alive by summoning up thoughts of my wife and the prospect of seeing her again and by dreaming at one point of lecturing after the war about the psychological lessons to be learned from the Auschwitz experience."

Stripped from his home and family, Frankl found himself on a train with 1500 persons towards an unknown destination. At one point during the journey, they passed a sign that read Auschwitz.

An uncomfortable silence filled the train car because they all knew that that name stood for all that was horrible; gas chambers, crematoriums, and massacres. The situation before him was dire, and it would only get worse. Once they stepped off the train, they were separated into groups on the left and right.

The significance of the sorting was figured out later in the evening when the survivors found out that it was the first verdict made on their existence or non-existence. Frankl highlights that during an initial couple of days, he and the other prisoners were hopeful that everything would work out for them until they were marched to the *disinfection* chamber.

They were forced to give up all of their earthly possessions, and they had not grasped the fact that everything would be taken away from them. Even if they had decent shoes, they were exchanged for shoes that did not fit.

Guards beat them mercilessly if they tried to deceive the guards, hiding possessions, or alter their shoes. "Thus, the illusions some of us still held were destroyed one by one, and then, quite unexpectedly, most of us were overcome by a grim sense of humor. We know that we had nothing to lose except our naked lives."

Frankl describes the camp as follows.

> There isn't a collection of words that could be put together to describe how truly awful this situation was. One thousand five hundred captives cooped in a shed built to accommodate 200 at the most, one five-ounce piece of bread and broth in a week, the genuine palpable fear of death at any moment, two blankets shared by nine men in freezing temperatures, unable to wash or brush your teeth, just to name only a fraction of the things they had to deal with.

Through this impossible situation, Frankl talks about a specific day the prisoners went to the worksite. While digging ditches in preparation for railroad tracks, one prisoner asked Frankl, "If our wives could see us now! I do hope they are better off in their camps and don't know what is happening to us."

This brought thoughts of Frankl's own wife to mind. Through miles of walking, slipping in icy conditions with inadequate shoes and clothes, supporting each other, and at times dragging each other along the path, silent contemplation of his wife engulfed his emotional state.

Frankle wrote in his book, "Occasionally, I looked at the sky, where the stars were fading, and the pink light of the morning was beginning to spread behind a dark bank of clouds. But my mind clung to my wife's image, imagining it with an uncanny acuteness. I heard her answer me and saw her smile and frank and encouraging look. Real or not, her look was then more luminous than the sun, which was beginning to rise."

In the middle of the most hopeless situation known to man, Frankl shifted his focus to what mattered the most, his wife. He was unaware whether or not his wife was even alive, but that didn't matter; he exercised his power of choice in his most significant moment of suffering. In the immense darkness of night, he chose to rest in the light of the stars.

His choice of perspective helped him to understand that "forces beyond your control can take away everything you possess except one thing, your freedom to choose how you will respond to the situation. You cannot control what happens to you, but you can always control what you will feel and do about what happens to you."

Frankl practiced intentional gratitude in the worst conditions and clarified what was most important—he discovered his *why* in the bleakest situations of life and focused on gratitude for that daily.

Intentional gratitude refers to a conscious and deliberate practice of focusing on and acknowledging things for which one is thankful. It involves reflecting on what one appreciates and expressing gratitude to oneself or others. The goal is to increase one's overall sense of well-being and happiness by cultivating an outlook that isn't overwhelmingly negative.

The benefits of intentional gratitude are overwhelming:

1) improved physical
2) emotional and social well-being
3) greater optimism
4) increased self-esteem
5) heightened energy levels

6) strengthened heart and immune system

7) decreased blood pressure

8) improved emotional intelligence

9) expanded capacity for forgiveness

10) improved self-care

11) greater perspective

Why does this matter as a leader? As the leader, you are the problem solver, you are the strategist, you are the solution bringer, and you are expected to figure it out by any means necessary. There are times when things look bleak for you and your team, and they are looking to you for the hope that things are going to be alright. You are expected to look at any situation, no matter how dark, and see possibilities. The fight in this round is the hopelessness accompanying the problems you are expected to fix.

Oftentimes your existence as a leader will be filled with navigating what is perceived as bad news. The truth is there is no such thing as *good* or *bad circumstances*. There are just circumstances that we are challenged to resolve. Leaders define everything that occurs as an opportunity.

The most outstanding leaders in history, Martin Luther King Jr, Nelson Mandela, Gandhi, and Churchill, are only a few examples of leaders who were tried by fire and came out on the other side to lead others safely through that same fire.

It is easy to see these people as the exception or as special. The truth is, they weren't very different from any of us. They were ordinary people who were presented with abnormally difficult situations. The actual difference lies in how these individuals saw those problematic situations. They interpreted these situations as opportunities instead of obstacles.

Amid hard times, human nature programs us to freeze, flight, or fight. Most of us submit to this automatic built-in response, but leaders must take this power back and change our experience of the world. We need to remember that contained within our response to hardship lies our ability to choose how we see, respond, and act. Nothing we experience has any meaning except the meaning we give it.

Leaders must fight every moment of the day to take ownership of that meaning. By taking control of the importance of his suffering, Victor connected with the *why*. His primary *why* was his wife and his family, and his secondary *why* was to determine whether his life's work in psychology held up during one of the darkest times in human history. His thoughts in his darkest moments were the unknown people that could be helped through his suffering.

Frankl said, "A man who becomes conscious of the responsibility he bears towards others, or to an unfinished work, will never be able to throw away his life. He knows the *why* for his existence and will be able to bear almost any *how*."

I can relate to this responsibility. In 2007, I took over as Leading Chief Petty Officer onboard the USS Boise SSN-764. This was my first tour of duty as the legitimate leader of my team, and I was *The Chief*. I was supremely excited about the opportunity to lead and had a preconceived idea about how things would go.

Everywhere I had served prior to the Boise, I had been successful, and I was not ready for what faced me when I showed up. I took over for a leader that hadn't been doing the job for the past seven months due to a personal issue.

The division needed direct guidance. During that time, all of our division's equipment was operating at a limited capacity. I was told that all the people in my division were below-average performers and that motivating them would be a struggle. I disavowed every opinion that others had about my team and met them where they were.

They hadn't been trained; even worse, they hadn't been valued. I knew this would be a considerable undertaking, and this was my first appointed leadership position. I was expected to have all the answers, and it seemed I had none for the first six months of my tour.

Our equipment was still not operating properly, my team was still learning, and I was taking the brunt of the criticism. It was a very dark time for me, and my positive outlook was diminishing daily. I knew the situation would worsen before it got better because everything needed to be broken down and built back up correctly. The training method needed to be revamped, and the relationships and culture of the team needed to be rebuilt.

Knowing it would get worse before better didn't stop me from occasionally thinking that the situation was hopeless and would end up with me fired. The small glimmers of light that shined through my team allowed me to maintain perspective during this dark period.

I was grateful for the incremental areas of growth that my division showed. I would celebrate those moments and probably go overboard, but I wanted them to know that they were valued above all.

I was grateful for each lightbulb moment they would have along the way. My personal leadership philosophy was to focus on helping people be the best version of themselves rather than being just good Sailors. That meant helping them to find their *why* despite their depression and hopelessness.

Once they had their *why* they needed to live in that gratitude daily. That is the formula, but the journey to achieve that is the true miracle of outstanding leadership.

For me, the primary area of strength came from the encouragement of my wife. When I descended into an emotional funk, she would pick me back up simply by reminding me about my *why*. I would have the opportunity to practice gratitude in my *why*, not only for myself but as a living example for my Sailors.

I desired, above all, to be the leader I wanted to be early in my career. She was with me during those hard times and would remind me that I was there to improve lives despite what those above me thought about what I should be doing. She was my primary source of intentional gratitude through that dark time.

One short year later, my division was not only the best on the boat but the best amongst all six of the other submarines within our squadron. For five of the six years I was the Chief; my crew celebrated a top division five times!

What had changed? Through my leadership, I helped them rebuild their hope that this was possible. It took trust, time, and tenacity, but they did it. They slowly took to heart the idea of *belief* instead of *hopelessness,* and now, they experience the joy of the outcome.

Where there is hopelessness, there is a decision to be made; you can decide to remain in that hopelessness or to find a belief—your *why.* As in Frankl's case, he had an opportunity in his darkest time to choose hope—he defined his *why*! Despite being stripped of everything, he found his *why* and stayed grateful for it!

Frankl's decision teaches leaders about the importance of having a clear sense of purpose in their work. Without that purpose, without that *why* our circumstances will control and direct our actions with a clear *why* leaders can win the fight and transcend any hopeless situation.

ROUND 10

THE CHALLENGER—INSECURITY

People buy into the leader before they buy into the vision.

—JOHN MAXWELL

"You are the absolute worst leader I have ever seen! Why did the Department Head ever think you would be the right choice? You don't have any idea how to talk to or treat people. If I have to deal with leaders like you for the rest of my career, I think I'll just get out!"

This is a direct quote from one of the Sailors I was charged to lead early on in my leadership journey. Like most leaders, we are thrust into leadership roles based on our performance at assigned tasks. Let's just say the first few months didn't go so well.

"You want me to do what, sir?" I asked, shocked.

My department head responded, "I want you to lead your team and replace the Chief. He isn't performing to standards, and you are the person I want leading the charge to make sure we are ready to fight."

"Sir," I stammered, "I am just a Second Class Petty Officer (E-5). He is two full ranks, fifteen years my senior, and I have no idea how to lead anyone. I have a hard enough time leading myself." I exclaimed.

"I don't care," he barked, "It is done. Figure it out. Petty Officer Owens, don't screw it up."

"Yes, sir," I said with everything I could muster.

I had never been more terrified in all my life. I had to take the place of a Chief Petty Officer (E-7) and lead a team that was directly responsible for the safe navigation of a submarine. A Chief Petty Officer in the Navy is the on-scene leader charged with ensuring all divisions/teams are adequately prepared to take the submarine into any situation and in harm's way if necessary.

Chiefs have received extensive leadership training and an average of ten years of experience. They are technical experts, and there are very few things that they can't repair. Chiefs are the true *GO-TO* individuals within the Navy that are the most resourceful, adaptive, and versatile leaders.

And here I come, a lowly E-5 with four years of experience and no clue how to lead a group of people. I had never been more overwhelmed. The stakes were extremely high. Should I fail, we would run the risk of hitting another submarine, losing our position underwater, hitting an underwater mountain, or losing all communication onboard, just to name a few of my failed leadership journey possibilities.

I had to get this right! I did not get this right! My first mistake was attempting to emulate the leaders I encountered along my journey. I screamed, yelled, and degraded my team and withheld information from my team; I was slow to make decisions and quick to change them along the way. Like trying to walk in shoes too big, all

I did was trip over myself and fell flat on my face because I attempted to do everything myself.

I went to work early and left late every day to learn all I could and make up for my lack of experience. Despite my external efforts to mitigate my lack, I was clueless about addressing what was happening inside of me. I leaned heavily on the technical documentation and subverted the experience and knowledge my team had already gained along the way.

My second mistake—I knew this was my opportunity to show and prove. I wanted nothing more than to come to the rescue and show everyone that I was a capable leader well ahead of the curve. I wanted people to be in awe of how I handled challenges seamlessly and without issue.

I wanted to show our Chief that I didn't need him and that the Department Head was right about me. Despite what I wanted to do, with every mistake I made along the way, my inner playlist consisted of the following:

"You're in over your head!"
"You're not supposed to be here!"
"Failure is imminent!"
"Everyone will know you're a fraud!"

I allowed this playlist to play on repeat day in and day out. I only focused on what was going wrong. It wasn't all bad, but the reality I crafted for myself with my own words was filled with failure. The outer result that I desired was incongruent with who I was at my foundation.

"And the rain descended, and the floods came, and the winds blew, and beat upon that house, and it fell: and great was the fall of it." (Mat 7:27 KJV). Inside I was insecure, shallow, scared, anxious, and timid. The opponent in this round is insecurity.

Truthfully, I wanted to project that I was "the man." I was unconcerned with the actual mission of the submarine. This was all about me and my ego. I wanted to be a leader without working to become an authentic leader.

My lowest point occurred as we prepared to get the submarine underway to go to sea for an essential operation. Getting a nuclear submarine underway is a very challenging and daunting exercise. Submarine operations are illogical; let's take this black tube full of people and submerge it on purpose without filling the *people's tank* with water. It is one of the most dangerous jobs on the planet, with very little room for error.

While getting the submarine ready, we received an error with our navigation system. Our navigation system is slightly different from the GPS unit on your phone. This system must be managed very carefully because it keeps our underwater position. Any mismanagement could be the difference of a few feet or miles, and that error level can lead to a catastrophic outcome.

One of my team members let me know about the error, and I told him to enter a specific code to make the error disappear. He hesitated at my order and said he wasn't sure it was the correct code. He began to leave to get the technical manual to make sure.

I was immediately furious. I perceived his actions as a slight, as though he was questioning my knowledge and skill. I told him the right answer, so why would he need to get the manual unless he thought I was incompetent? My insecurity crafted an entire narrative just because he wanted to do the right thing. Angrily, I stopped him and told him to enter the code; there was no need to verify it.

A short while later, while focusing on the next task to get us out to sea, all of our navigation system alarms began going off. We all rushed into the control center to assess the problem, and the petty officer who entered the code was standing at the navigation system, looking terrified.

"What happened?" I asked.

"I entered the code," said the petty officer, "as you told me, and the system entered a 24 hour calibration mode."

"Oh, my God," I shouted. "I'm going to get fired. I must tell the Department Head and the Captain that we can't get to sea today because I screwed up our navigation system."

Terrified, I left the control center and let our department head know the status of our systems. I told him that I had given the order to enter the code into the system without verifying it and that it was all my fault. The system entered calibration mode, which no longer provided reliable position information. I was completely devastated.

As I relayed this information, he yelled, yelled, and then yelled some more. He told me I had let him down and that it was a mistake for him to put me in charge. His words were a confirmation of how I felt about myself inside. We went to brief the Captain, who acknowledged the information and told me to step out of the room. He proceeded to *lay* into the department head for putting me in charge.

I left the submarine to go topside and get some air. I have never been more broken. I sat topside with my head in my hands, clueless about proceeding forward. I couldn't move or speak and wanted nothing more than to disappear. No crew member talked to me while I was near tears sitting on the pier outside the submarine. After a few minutes, I could feel someone standing over me, but I didn't dare to face anyone.

"You're just having a kintsugi moment." someone said.

"What?" I said, lifting my head to see who was talking to me.

"Yeah, this is great for this to happen to you this early. Your kintsugi moment," he replied.

"Great? Great? I am single-handedly responsible for this warship not getting underway to do its job. Nothing about this is good." I muttered.

"Yeah, I know you don't see it now, but this will be one of the greatest things ever to happen to you. You're broken now. You believe you're beyond repair, but I've heard about your potential and ability. You'll be just fine. Before reporting here, I spent time in Japan and learned about their philosophies on life, specifically kintsugi. When you get home today, you should look it up. It gets better."

He was a Lieutenant (0-3) division officer from another submarine across the pier. I didn't even know who he was, but I took his advice and looked up kintsugi when I got home that night. After several minutes of trying to spell kintsugi correctly, I finally learned what the LT was speaking about.

In the late 13th century, the shogun or commander in chief for Japan broke his favorite bowl and sent it away to be repaired. Eager to have his bowl back, he was supremely disappointed in the quality of the repairs, as it was put back together with metal staples. He then charged his team to come up with a better solution.

Instead of attempting to fair in or hide the flaws, they highlighted the damage to the bowl by using the flaws to make it more beautiful.

Kintsugi is a two-part word with kin meaning *golden* and tsugi meaning *to join together.* It literally means to join together with gold. Kintsugi is a principle of Zen that states that pots, cups, and bowls that have become damaged shouldn't be cast out or thrown away. They should be carefully reassembled and glued together with lush gold powder and lacquer. The goal is not to hide the damage but to make the fault lines beautiful and strong.

The traditional kintsugi method begins with the gold lacquer to glue the pieces back together. The lacquer is also used like putty to fill in gaps or holes where fragments from the original may be missing.

According to kintsugi experts, mending is the most challenging part because the lacquer cannot be removed once it's dry, and the pieces must be put into place simultaneously, regardless of the number of different parts.

The greater the degree of brokenness, the greater the potential for beauty.

After spending all night digesting the reality of my brokenness and the kind of leader I could become, I realized that I was my own worst enemy. The chance interaction with the LT helped me know that I was the one orchestrating my own miserable symphony. Looking at my circumstance through the lens of Kintsugi, I saw myself as a vessel that was initially constructed weakly and couldn't hold the weight of the burden of leadership.

I could see how I needed to put myself back together through the brokenness. I had torn myself down with my self-talk, and if I was ever going to be the leader I desired to be, I needed to build myself back up with my self-talk.

I realized that the lacquer I was to be forged with must be words of encouragement, not criticism. Through the wisdom of a broken bowl, I saw that the expectation of perfection that I had for myself was delusional and impractical. If I was going to change the narrative in my head, I had to forgive myself for not being perfect.

No wonder we've become so obsessed with seeking perfection. Antonyms for *perfect* are the following words: Flawed, corrupt, inferior, poor, second-rate, inept, broken, wrong, evil, and all negative. Who wants to be considered inept? But the truth is that the benefit of imperfections allows us to connect with the people we have the privilege of leading. We are all imperfect, and what we do as a result of that awareness is where we make our marks as leaders.

At that moment, I started to construct my own personal restorative lacquer. First, I began to write down everything that was going well and what I did well. I needed to silence the discourse of failure because that was all I could focus on. So I chose a different focus point. Remember, regardless of how bad things are going, it's not all bad; it's merely a matter of perspective.

To change the narrative, I had first to change my perspective. I wrote down all of the good and always kept it with me. Whenever I felt a cloud of defeat creeping in, I would pull my sheet out and allow the truth of who I was to dissipate those clouds.

Next, I began to be kind to myself and give myself grace. I decided to give myself the benefit of the doubt for not being perfect. I would double down on the things I did well and collaborate with my team on the things I didn't. The goal was mission accomplishment, and it doesn't matter who gets the credit.

These two ingredients became the restorative lacquer I used to put myself back together. I knew it wouldn't be easy, but it would be worth it. My team did not deserve a leader who cast a shadow over their lives because the leader was incapable of confronting their inner mess.

I gathered my team together first thing the following day. I acknowledged my shortcomings and my faults. I apologized for my approach and my mistrust. I told them I was more interested in looking good than doing good. I was transparent and honest about how I felt about my job and why. I told them I would do better because they deserved the best possible leadership.

I then went to speak with my department head, who was still fuming from the day prior. I told him he didn't make a mistake putting me in charge and that I would be better from here on out. I told him that I didn't subscribe to quitting just because things got hard and that it was on me to get harder. He laughed and told me to get back to work to get the submarine underway.

After about a year and a half, that Sailor who said, "I was the worst leader he'd ever seen," pulled me aside and told me he appreciated my effort to improve. He apologized for what he said to me and thanked me for being authentic and genuine.

He told me he had learned more from me in the last two years and believed he could be a great leader because of my approach. I was humbled by his words and took out my paper and wrote that down.

It took some time to get the lacquer to the right consistency, but once it did, I learned how to leverage every mistake, misstep, and failure into growth and value on my part and, more importantly, part of my team. I no longer pursued perfection but excellence.

I never lose. I either win or learn
—NELSON MANDELA.

ROUND 11

THE CHALLENGER—SELF-AWARENESS

*To be yourself in a world that is constantly trying to
make you something else is the greatest accomplishment.*

—RALPH WALDO EMERSON

Yes….yes…..yes….
It rings out—the chant of thousands of people on their way to conquer their Goliath.

Yes….Yes……yes…..

12:12 in the middle of the night, the only illumination coming from 1000-degree hot coals.

Yes….Yes…..Yes….

We've been preparing the entire day for this moment.

YES….Yes…..Yes…..

I step up to the platform, hot coals before me, barefoot, determined, and focused.

YES….YES….YES….

I get a signal, make my move and go…

Yes.

Every day, leaders are challenged to navigate uncertainty, issues, and challenges with ease, calm, and a definiteness of spirit. The people we lead expect us to have the right answers in every scenario, regardless of difficulty.

When we are appointed to a leadership role, the primary determining factor for our readiness for that role is our experience. Have we shown a positive trajectory in our performance that speaks to our potential as a leader?

The truth is experience is rarely a great indicator of performance in a leadership position. Even though science has proven that promotion based on previous experience is flawed, most organizations continue to promote this way. This philosophy is known as the *Peter Principle.*

The *Peter Principle* is a well-known concept in management and leadership that states that employees within a hierarchical organization tend to be promoted to their *level of incompetence.* In other words, employees will continue to be promoted until they reach a role they cannot perform effectively, at which point they will no longer be promoted.

This can lead to a situation where individuals are placed in leadership positions for which they need to be qualified, leading to poor performance and potentially negative consequences for the organization.

Dr. Laurence J. Peter first proposed the Peter principle in his 1969 book *The Peter Principle: Why Things Always Go Wrong.* Dr. Peter, a Canadian-American organizational psychologist, argued that organizations often promote employees based on their past performance rather than their potential or suitability for a particular role. This can lead to a situation where individuals are promoted beyond their abilities, leading to a lack of effectiveness and potentially even harm to the organization.

Why does this matter to you? Well, the premise of this book is that leadership is an inside game. Leaders must take responsibility for their competence, character, and overall readiness for their appointed position. Despite the organization's inability to recognize that their promotion tactics are outdated, that doesn't remove our individual accountability to be ready.

The fight in this round is our own level of self-awareness.

Are you legitimately ready for the next position of leadership? If not, what things do you need to do to become ready?

Self-awareness is the ability to recognize and understand one's thoughts, feelings, and emotions and their impact on others. In contrast, it may seem like a simple concept, but developing and maintaining self-awareness can be challenging for many people.

One reason that self-awareness can be difficult is that it requires introspection and reflection. This involves looking inward and examining one's own thoughts, feelings, and behaviors, which can be difficult and uncomfortable for some people. It may be especially challenging for those who are used to externalizing their problems or have difficulty acknowledging their own emotions.

Another reason that self-awareness can be difficult is that it requires the ability to be objective and unbiased. This means seeing oneself and one's actions from an outsider's perspective rather than being overly influenced by ego or personal biases. This can be difficult for anyone, as we all naturally see ourselves in a positive light and justify our actions. We all have a tendency to get caught up in our heads and fail to see ourselves through an honest and realistic lens.

So there I was the Command Master Chief of Pearl Harbor Naval Shipyard. I had recently completed a successful tour onboard the USS Helena as Chief of the Boat. My career was at a major turning point because I didn't get the job I thought I should've gotten after the USS Helena.

I believed my performance was good enough to warrant my selection to what I perceived to be a higher position. Even though I was really happy to be in Pearl Harbor and serve the Sailors and civilians, I thought it wasn't where I was supposed to be.

In my anger and frustration about not getting the job I wanted, I became supremely defeatist and reasoned that my career was over in my current field. I internalized all of these feelings and saw myself through a distorted view, and though I wasn't depressed at home, I garnered no job satisfaction and blamed those appointed over me for not seeing my value.

Sensing my dismay in one of our conversations, one of my good friends recommended that we go to a conference called *Unleash the Power Within* by Tony Robbins. Tony Robbins is among the most well-known and successful motivational speakers and life coaches.

Early on, I bought into the rumors that people would tell me about Tony Robbins. "Dude, you're not going to spend that much money to see the dude on the infomercials, are you?" Uhhhh, yeah. I'm going to give it a chance, but I promise I won't drink the kool-aid.

Plot Twist: I drank the kool-aid.

The experience took me a whole year to digest and understand. Every moment of the conference, no, the experience, was thoughtful and purposeful. As soon as you walk into the door, you are hit with a wave of positive energy that becomes intoxicating.

You've never seen so many people smiling, jumping, and singing in anticipation of the person they will become by the end of the experience. Everyone knows. They feel it intuitively. Instantly, you know you will never be the same person at the end of the experience.

I'd often heard people who have experienced life-changing moments that made them feel like they had been reborn. As I walk through the conference room, I see a huge picture of a butterfly, and I understand immediately.

The transformation into the butterfly only happens once. The butterfly only gets one transformation. We can re-enter the chrysalis and emerge even more glorious and majestic than before. We aren't forced to remain the person we are despite our circumstances. We can't control the environment, but we can control ourselves and our response to what is occurring.

Our transformation should be perpetual. I immediately thought about the honor of being a leader. People trusted me to lead them to success. Personal success, organizational success, and family success. The people I led deserved the best that I could possibly give them. They deserved the best version of me.

How could I give them the best version of myself if I didn't know what I could do? How could I be my best if I didn't know what my best was? As a leader, I felt that I owed them a personal dedication to challenge myself to re-enter the chrysalis and emerge better than I was yesterday.

At the conference, my friend and I agreed that we would *play full-out* and maximize every minute of the experience. We jumped with, danced with, high-fived, laughed with, and cried with everyone in our purview. The day's message was clear: YOU NEED TO KNOW WHAT YOU ARE CAPABLE OF.

Now I had heard of the rumors about what was to occur on the first day. People whispered about walking on fire, hot coals, or burning leaves. I thought this was an exaggeration. Surely, the liability associated with the possible injuries would prevent someone from doing this for real.

Indeed this was a psychological trick that would give you the sense of walking on fire rather than actually walking on fire. I had already reasoned in my mind that this wasn't a thing that was going to happen. This would certainly be smoke and mirrors, and I wondered how they would pull that off.

Firewalking is an ancient ritual that has existed for thousands of years. Practiced by different people around the world, it has roots in many different cultures. The common thread that all firewalking rituals seem to share is that the firewalk itself demonstrates courage, faith, and strength—the ability to stand up to one's fears and take on whatever challenges life brings.

The earliest documented firewalking rituals date back to 1200 BC, when the first recorded firewalk took place in India. Two Brahmin priests took the firewalk together as a competition; the priest who walked further had this feat recorded. Even then, the firewalk was clearly a metaphor for spiritual strength and calmness of mind.

Firewalking is often used in healing ceremonies. The 17th-century Jesuit priest Father Le Jeune once witnessed a healing firewalk among the North American Indians. In a letter he wrote to his superior, he described seeing a sick woman walking through fires with bare feet and legs. She was not burned, and she even stated she felt no heat at all.

The Bushmen of the Kalahari desert have always used fire and firewalking in their powerful healing ceremonies. Anthropologist Laurens van der Post observed firewalking as part of their healing rituals and went on to write about the practice. Other cultures worldwide have used firewalking rituals for initiation, to mete out justice, as displays of faith, and for many other reasons.

Firewalking has been practiced in various countries in South Asia, Africa, Europe, Central Asia, the Caribbean, East Asia, the Pacific Islands, Polynesia, the Mediterranean, and North America.

Firewalking falls into two basic types. One of these is walking over hot stones or other objects, a form prevalent in Polynesia. The other and much more widespread form is walking over live coals. The fuel em-ployed in the latter is usually wood, pref-erably a kind that will produce red embers retainable for a considerable time. The walk does not begin until the fire is white-hot, with blue gaseous flames flickering an inch or two above the surface.

Though I was certain I wouldn't be walking on actual hot coals, the experience process placed every reason I clung to, justifying my fears, being placed before me, and utterly destroyed me. So at the end of the night, I am convinced without a shadow of a doubt that I can do the impossible.

Through the experience, you become very aware that you have been living life at a six, convincing yourself it's the best that you can do. You get a glimpse of who you can be at ten, and you know who you were eight hours prior no longer exists.

I've never been so tired and energized at the same time. Never have I been so cold and on fire—mainly because Tony keeps the arena at about 55 degrees.

I came in with my own expectations. I traded them for appreci-ation: appreciation for the moment, appreciation for the people I get the honor of spending this time with, and appreciation for the op-portunity to bring the best version of myself back to the people who trust me.

I constructed a blueprint of how I felt this experience should play out through my expectations. It wasn't until I forsook my blueprint and allowed myself just to experience each moment that I truly un-derstood.

We have such a delusional idea of control. We subconsciously believe that we can control everything external to us. We spend so much time attempting to conform the world to our desires. The trick is that we can only control one thing; ourselves.

Instead of spending so much time attempting to control the ex-ternal, we should expend that energy controlling ourselves. Specifi-cally, controlling our response to the hand that life deals us. My thoughts floated back to the circumstances surrounding my assign-ment in the Navy.

I realized I became entitled and resentful because I didn't get what I felt I owed. The truth was, I wasn't owed anything, but I owed everything if I counted myself among those considered leaders. I was never promised the position that I wanted so dearly, but if I re-mained in my current state, I would never progress to another posi-tion of prominence.

I had to be willing to let go of what I thought were certainties for what was happening now and who I was now. I viewed myself through a self-serving lens that almost imprisoned me in jail or should have and could have.

To be the best version of myself for the people I lead, I had to be honest with myself and determine how I would improve because I was still in progress.

As we approached the culminating event, collectively, we all believed that we could do the impossible. We believed that there was a portion of ourselves that we hadn't fully tapped into. The staff directed us to begin the walk outside. As soon as we got to the door, the smell of fiery embers filled the air. Contrasted against the darkness was the bright red glow from searing coals, and you could feel the nearly unbearable heat with which we were to do battle.

Those coals represented my false projection of myself. Those coals represented the fabricated persona of who I thought I was. If I was to become the leader my people needed me to be, I needed to challenge my limitations. By challenging my limitations, I could see myself through a clearer lens.

When the pressure of this confrontation begins to take its toll, the weaknesses, the fears, and the insecurities come to the surface in the form of excuses.

"You're doing fine; you don't need to go this extra mile."

"This is too much; no one else is challenging themselves in this way."

"What are you trying to prove?"

When we intentionally take a step to reach the next level, the dialogue in our minds will always attempt to convince us to retreat to safety. Every instinct within us will direct us toward safety.

This is how our comfort zone becomes a prison, and we become blinded to the chains of our own construction. It is a painful process to confront our limitations. It is a painful fight to increase our self-awareness because you have to admit what you are NOT to become who you need to be.

The question is, *is the desire to be the best for your people stronger than the fear of the pain of growth?* Growth and maturity in this area do not come naturally. We don't inherently become more self-aware unless we intentionally pursue that goal.

As I approached the fiery coals, I could hear every excuse in the book not to proceed further. I could hear the *ME* in my comfort zone arguing with the *ME* desiring to break free. The distance was a mere seven or eight feet, but it initially looked like a mile. Gathering myself, I thought about who I would be on the other side and repeated the mantra we were taught…

YES…one step…
YES…two steps….
YES…three steps….
YES…four steps…
YES…nine steps…..
YES….10 steps…
YES!

I. Walked. On. Fire. Not a blister, burn, or scar—nothing. I conformed myself to the situation. So in control that the burning coals didn't burn me.

I was incapable of putting into words the level of elation that came over me at that moment. To defy the very nature of the world we live in simply because I made a choice to believe that I could be the equivalent of emerging from the chrysalis anew.

As leaders, we have to fight against our perceived abilities every day. We fight with ourselves about what value we can bring to our teams. We fight with ourselves about whether or not we can achieve what has previously been deemed impossible.

To overcome this fight, we must continuously challenge ourselves to grow to higher heights. We must continuously run head-first into what we fear and defeat its power over us. You don't have to walk on fire to achieve the results you desire necessarily.

Through these individual feets of accomplishments, we set the tone through our unwavering demeanor for our teams and infuse them with the same level of faith and confidence we experienced. By setting ourselves ablaze, we become a light to their paths.

By intentionally taking on new challenges, you will unlock the capability you need to become the leader you desire to be and, more importantly, who your people need. The self-awareness gained during these challenges release a greater sense of self-efficacy and self-confidence.

You will gain new perspectives and insights that allow you to see with clarity no matter the situation. Dr. Robert Holden, the founder of *The Happiness Project*, said, "The more you know yourself, the less you are likely to get lost in the chaos of life."

Leadership is not just about making decisions and giving directions; it's also about understanding one's strengths and weaknesses and adapting to different situations. A self-aware leader can better understand their own biases, thought processes, and triggers, enabling them to make more informed decisions and lead with greater empathy and understanding.

ROUND 12

THE CHALLENGER—LIVELIHOOD

Anyone can hold the helm when the sea is calm.

—ANONYMOUS

He withdrew from them about a stone's throw, and He knelt down and prayed,

Father, if You are willing, remove this cup from Me. Nevertheless, not My will, but Yours, be done.

An angel from heaven appeared to Him, strengthening Him.

And being in anguish, He prayed more earnestly. And His sweat became like great drops of blood falling down to the ground. Luke 22:41-44 (KJV)

I n *The Fight to Lead*, every chapter and previous round leads to this penultimate conclusion—the final round. The final round is known as the *Championship Round* because that's where champions are made.

The eleven previous rounds have prepared you for the real battles you will face as a leader—from overcoming insecurity in your environment and navigating past hurts to squashing your ego. The final opponent for this round is your own livelihood—the complete and utter denial of self to ensure the prosperity and, in some cases, the survival of those you have been appointed to lead.

The ultimate goal of leadership is to serve the people we have the privilege of leading. It's about *their* success, not the leader. All the leaders' efforts and energy are directed toward the prosperity and advancement of those under their care.

This naturally means that the leader must put themselves on the back burner. Sometimes, the leader must be willing to sacrifice their own well-being for the team's benefit. It's not an accident that the captain is expected to *go down with the ship*. Leadership, at its core, is about service, and there is no separating service and sacrifice.

Here lies the reason for the *lack* of genuinely impactful leaders worldwide. It is impossible to indeed be of service without having to endure sacrifice. Sacrifice is the true fight of leadership, and it is the burden of the leader to meet the needs of his or her team so that they may fulfill their potential.

That burden may ultimately result in your literal or figurative death. Figuratively, you die to yourself and what you hold to be true to serve your people. To *die to oneself* means to put aside one's own ego, desires, and needs to serve and uplift others. *Dying to oneself* is the willingness to confront what you know and believe and abandon those perspectives to help those you lead.

Becoming the leader our people need is the process of subtraction and not addition. Regardless of the number of books, courses, or seminars, one adds to themselves, if the foundation of the individual hasn't been cultivated with service, great will be the fall of that person when the storm comes.

It requires a selfless attitude and a willingness to prioritize the well-being and success of others over one's own. This can involve making sacrifices and setting aside personal biases to understand better and meet the needs of those being served.

In a leadership context, *dying to oneself* means putting the needs of the team or organization ahead of one's own and being willing to make sacrifices for the benefit of the group. This type of selflessness and service takes you from an appointed to a transformational leader.

In my line of work, you may be required to make the ultimate sacrifice for your people, like Marine Pfc Jack Lucas, a United States Marine who served in World War II. Jack was the youngest person to receive the United States military's highest decoration for valor, the Medal of Honor, at 14.

Lucas enlisted in the Marine Corps at 14, lying about his age. He was sent to fight in the Battle of Iwo Jima, where he was involved in hand-to-hand combat and saved the lives of several of his comrades by throwing himself on top of two enemy-thrown grenades, absorbing the blast and saving the lives of his fellow Marines.

Like Army Spc John Baca, who served valiantly during the Vietnam War, he joined the Army in 1968 and served with Company B, 2nd Battalion. He was awarded the Medal of Honor for his actions on April 6, 1970, near Firebase Blanca, Vietnam. Baca, who was a specialist and a team leader, saved the lives of several of his comrades by throwing himself on top of a live grenade during an enemy attack, absorbing the blast and saving the lives of his fellow soldiers.

Like Marine Cpl Duane Dewey, who joined the Marine Corps in 1951 during the Korean War, Corporal Dewey earned the Medal of Honor on April 16, 1952, near Panmunjom, while serving as leader of a machine gun squad with Company E, 5th Marines, 1st Marine Division. He had already been wounded by a grenade that had exploded at his feet and was being treated by a navy medical corpsman when another enemy grenade landed at the squad's position.

Yanking the corpsman to the ground and warning members of the squad, Dewey flung himself on the grenade shouting, "Doc, I got it in my hip pocket!" The grenade exploded, lifting Dewey off the ground and inflicting "gaping shrapnel wounds throughout the lower part of his body." In addition, he sustained a bullet wound to the stomach.

These men understood the fight to overcome the survival instinct to preserve themselves versus the necessity of preserving the ones they lead. *Nevertheless, not my will...*

Enter Jesus. Many perspectives exist about who Jesus was, and I am not here to debate that in this text. What is undebatable is how those who encountered him, wrote about him and ministered with him felt about him.

Ultimately, the measure of the leader doesn't rest with the opinion of the one who leads but with those he or she has led. He called many things, prophet, king, teacher, and Lord, amongst others, but one thing is clear he was the *consummate leader*. The accurate picture of who he was as a leader was depicted in the text Luke chapter 22.

Regardless of what you may personally believe, Jesus' journey to the cross is a master class of leadership that any leader can learn from.

Jesus' ministry was all about doing good in every circumstance he found himself in and introducing a new way of life to those who would listen. His perspective conflicted with the established religious leaders of that day, and their answer to fighting his perspective was to kill him. Jesus knew that for his mission to be complete, he must sacrifice himself for the people he was sent here to serve.

The story picks up during *The Last Supper*, where Jesus shocks his disciples by revealing that one of them will betray him. Sometimes, despite your best efforts, no matter what you do, some may not respond to your leadership.

Having spent the better part of three years in Jesus' inner circle witnessing the miracles, the amazing display of power, and his ability to lead, Judas still denied him. You can be the greatest leader in the world, and some still won't get it.

In Matthew 26:14, Judas is witnessed going to the chief priests with the greatest level of dissent for Jesus to make a deal to turn him over to the authorities. As leaders, not only must we be guarded against serving our own self-interests, but we must also be aware that those we lead are fighting a similar battle as well.

Judas agrees to turn Jesus over for the payment of thirty pieces of silver. The silver he acquired was worth about $600 in today's valuation. A small sum of money compared to the value that Jesus would add to Judas' life and those he came in contact with. There will be those you pour into and give all you can to see them succeed that will reject your efforts.

Despite the possibility that some may not accept your instruction, they are still worth your service. Even though you may see their value, they may not see their own value, and you may be the one who plants the seed but never sees the harvest.

The farmers till the ground, providing the seed access to sunlight. Then, he waters the grounds and removes the weeds, all with no assurance of a harvest. This is the life of the leader. Sometimes we experience our own Judas, but that can't stop us from giving all we have.

During the dinner, Jesus is very aware that this meal would be his last, as illustrated in Luke 22:15 (MES). When it was time, he sat down, all the apostles with him, and said, "You've no idea how much I have looked forward to eating this Passover meal with you before I enter my time of suffering. It's the last one I'll eat until we all eat it together in the kingdom of God."

Jesus knew that suffering was imminent, and he responded by having a meal with his team and continuing to add value to their lives. Interestingly enough, even Jesus knew that the suffering was incoming, and he knew who was responsible for it, but he never took action to stop it.

When faced with hard choices or decisions, there will be opportunities to off-ramp to a more accessible or less complicated path, but the question is this; what choice will add the most value to the people you have the privilege of leading? What are we willing to give to see the advancement of our people?

No one understood this choice more than Malala Yousafzai—a 15-year-old human rights activist from Pakistan. Blessed with courage and determination at a young age, when her school was closed by the Taliban, at age eleven, she gave her first speech on the school closings directly opposing the Taliban.

Malala rose to fame and prominence for her advocacy for girls' education. She was leading the fight against tyranny and discrimination, inspiring people worldwide. Then in 2012, at the age of fifteen, she was shot by a member of the Taliban while on her way home from school in an assassination attempt. This would have been the perfect opportunity to stop her advocacy.

Malala was only fifteen years old, and her life was in danger. No one would have blamed her for stopping; she could have chosen a less complicated path that didn't involve almost getting killed. She decided to focus on those without a voice!

> I raise up my voice- not so I can shout, but so that those without a voice can be heard.

> I don't want to be remembered as the girl who was shot. I want to be remembered as the girl who stood up.

> I am not a lone voice—I am many. I am Malala, but I am also Shazia, Kainat, Kainat Riaz, Shazia Ramzan, and Aisha. I am sixteen, but I am also every girl who has been shot by the Taliban.

Despite her injuries, she survived and continued her activism. In 2014, at the age of 17, she became the youngest person ever to receive the Nobel Peace Prize for her work promoting education and equality for girls. She inspires millions worldwide because she chose the more complicated path. Nevertheless, not my will.

After Jesus revealed to the disciples that one amongst them would betray him, they immediately started to bicker and point fingers at each other about who would be the culprit. This conversation quickly shifts to who would be the greatest among them. The life of a leader is problematic because, despite the value you add to their lives, you must always contend with one's desire to serve yourself.

While they were arguing, Jesus intervened:

> Kings like to throw their weight around, and people in authority like to give themselves fancy titles. It's not going to be that way with you. Let the senior among you become like the junior; let the leader act the part of the servant. Who would you rather be; the one who eats the dinner or the one who serves the dinner? You'd rather eat and be done, right? But I've taken my place among you as the one who serves. Luke 22:24-27 (MES).

As indeed the greatest the world has ever seen, he intentionally assumed the posture of a servant. The response from many of those holding leadership positions when faced with the idea of service is confusion and rejection.

When the term service is brought up, we think of lowly positions, janitors, maids, etc. We consider ourselves better than that; we believe we are above that characterization.

It is this posture that eliminates us from being the leader that our people need. If it weren't for those who choose to serve, the foundation that we count on would fall apart. Jesus' proximity to his ultimate sacrifice caused him to reveal his most important lesson to his followers.

After the conversation about who would be the greatest, he speaks directly to Simon. He tells Simon,

> Simon, stay on your toes. Satan has tried his best to separate all of you from me, like chaff from the wheat. Simon, I've prayed for you in particular that you not give in or give out. When you have come through the testing time, turn to your companions and give them a fresh start. Luke 22: 31-33 (MES).

Simon was confident that he would not give in and was ready for death if necessary. Jesus broke the hard truth to Simon that he would, in fact, deny Jesus three times when challenged by others about knowing Jesus.

When challenged with the hardest choice of his life, Simon failed under immense pressure and made the wrong decision three times over.

Though Simon, called Peter, was able to redeem himself later after Jesus' death, we often get one shot to make the right call during complex challenges. Leaders are challenged nearly daily to make the right decision for the people they serve.

Leaders like Winston Churchill during World War II had the most challenging task of keeping the people focused during great chaos and yet hopeful in destruction. As the Prime Minister of the United Kingdom during World War II, Winston Churchill had to make many difficult choices that would significantly impact the war's outcome.

One of the most significant decisions he had to make was whether to fight against Nazi Germany or make peace with them. At the time, many in the British government and the public were advocating for peace negotiations with Germany, believing that the war was unwinnable and that a negotiated peace would be the best outcome for the UK.

Churchill believed that making peace with Nazi Germany would betray the country's values and lead to a disastrous outcome for the world. He argued that the only way to defeat Hitler was to fight him and his regime.

This decision meant that Churchill had to take a stand against those who advocated for peace and to convince the British people and government to continue the war effort.

The hard choices for Churchill didn't stop there. He had to convince his people to continue fighting even when the war seemed lost. During the early years of the war, the UK faced numerous defeats, and the prospects for victory looked grim. The German army had conquered much of Europe and was closing in on the UK. The situation was dire, and many doubted that the UK could win the war.

However, Churchill refused to accept defeat, and he urged the British people to keep fighting. He gave speeches, rallies, and radio addresses to boost the people's morale. He convinced the people that there was still hope and that victory was possible.

Churchill's steadfast determination through all the adversity and naysayers was one of the primary reasons the good guys won World War II. *It is often through our most extraordinary times of suffering that the leader within is revealed.*

Nevertheless, not my will.

Through his most significant time of suffering, Jesus shows all leaders the ultimate blueprint for success. He illustrates, evaluates, and demonstrates.

Jesus always ensured his team knew his expectations in nearly every scenario. He didn't just tell them all at once and assumed they understood; he illustrated the principles he wanted to instill through the use of parables. Parables are simple stories that are used to reveal a moral or spiritual lesson.

One parable that Jesus told that is often related to leadership is the parable of the Good Shepherd in John 10:11-16 (MES):

I am the good shepherd. The good shepherd lays down his life for the sheep.

The hired hand is not the shepherd and does not own the sheep. So when he sees the wolf coming, he abandons the sheep and runs away. Then the wolf attacks the flock and scatters it. The man runs away because he is a hired hand and cares nothing for the sheep.

I am the good shepherd; I know my sheep, and my sheep know me—just as the Father knows me and I know the Father—and I lay down my life for the sheep. I have other sheep that are not of this sheep pen. I must bring them also. They, too, will listen to my voice, and there shall be one flock and one shepherd.

In this parable, Jesus compares himself to a good shepherd who cares for and protects his sheep, while a hireling (or a false shepherd) only cares for himself and will run away when danger comes. This parable illustrates the fundamental principles of good leadership.

Like a good shepherd, a good leader is someone willing to give their life for their followers and will go to great lengths to protect and care for them. They are selfless, and their primary concern is the well-being of their followers. They lead by example, and they're not afraid to put themselves in danger for the sake of others.

This parable also illustrates the importance of authenticity in leadership. A false leader, like a hireling, is only interested in their own gain and will not honestly care for their followers. Jesus is pointing out the difference between true leaders who are committed to their followers and those who are only in it for their own benefit.

Stories translate boring data into compelling pictures that have a better chance of being understood. He knew these stories related well to their environment so that they could understand the underlying message.

After extensively illustrating his expectations, Jesus would evaluate his disciples' performance. The primary tenets of Jesus' ministry were love, compassion, and service. With every teaching, he communicated these principles, and as they journeyed, he evaluated his team through the lens of those tenets. Jesus had a clear vision and expected his team to act accordingly.

He would numerous times challenge his disciples to demonstrate what he had taught them. He put them in situations where they could act autonomously through those principles and do anything toward those efforts if they worked through love, compassion, and service—caring for the sick, feeding the hungry, and standing against rules and regulations that acted against those principles.

Finally, he demonstrated the principles every single day. He was led by the principles that he demanded of those he led. No matter the challenge or obstacle, he acted on behalf of the ones he served. This demonstration is fully realized in the Garden of Gethsemane right before Judas betrays him.

After dinner and fellowship with his disciples, Jesus, with his core leaders, goes to the Mount of Olives to pray and prepare for his impending sacrifice. Significantly, Jesus would go to the Mount of Olives, considering the process of extracting the valuable oil from the olives through extreme pressing and his own individual pressing he was to endure.

The process of pressing olives to extract oil is a long and labor-intensive process that requires patience, determination, and a willingness to persevere through challenges. It also produces something of great value: olive oil used for cooking, skincare, and hair care.

In the same way, leadership is a long and challenging journey that requires patience, determination, and a willingness to persevere through difficult times. It's also a process that produces something of great value, a team that can accomplish great things.

Jesus always knew that his ultimate sacrifice was necessary to fulfill his service mission. As the time drew closer to that sacrifice, we saw Jesus alone, in the middle of the night, praying to God about another way. "Father, if You are willing, remove this cup from Me."

When faced with decisions of this magnitude, our normal survival response is to look for a way out. It is our inherent nature to seek an alternate path, ingrained within our nervous system, to protect ourselves.

When it is darkest for us when it seems as though we are all alone, we must take our eyes off of ourselves and consider the alternative if we do not follow through.

What is the cost if I don't step up? Who gets hurt if I fail to be courageous? What are the long-term repercussions of my unwillingness to do what was necessary for the people I have the privilege of serving?

On Dec 4th, 2019, one of the worst events in the history of Pearl Navy Shipyard occurred. An active shooter claimed the lives of two shipyard workers and injured another. An absolutely tragic and devastating event; it could have been so much worse.

I was the Command Master Chief of the Shipyard during this time and was in my office when this event was taking place. Outside my office, I heard police sirens near our location, and I went into the front office to see what was happening.

The executive leadership meeting had just finished, with all the shipyard leaders coming out along with the emergency management coordinator. He had a walkie-talkie and radio, and the information being relayed was too much to decipher exactly what was occurring.

Through all of the noise, I could hear something that sounded a little like "shots fired," but it wasn't clear, and the report, if it were shots fired, should have been "active shooter." As I was about to relay this to the coordinator, he passed over the radio, "fire, at drydock two."

The dread that swept over me as he spoke over the radio was sickening. No one else in the room even realized that he had just sent his emergency personnel into the line of fire. I immediately yelled at him that the report was "shots fired" and told his emergency personnel to shelter in place.

He looked at me confused, so I yelled again, "the report was *shots fired,* not *fire.* You are sending our people into an active shooter situation." He gets on his radio to make the correct report and diverts the emergency management team to shelter in place.

The truth is, I wasn't 100 percent sure that the report was *shots fired,* and I had to trust my intuition and gut feeling to make such a declaration that could have been wrong, delaying a response to a possible fire on a submarine. This is the struggle; this is the dilemma in leadership.

Are you willing to be wrong in the cause of doing the right thing? Are you ready to be courageous in a room full of senior leaders when no one else hears what you heard?

Afterward, I spoke with the emergency management team on the ground, and they received the call that it wasn't a fire at just the right time, and they were only seconds away from being directly in the line of fire of the active shooter.

We must be courageous during this fight to remove ourselves from the center of the universe. We must be strong enough to put aside our desires, ego, livelihood, and wants and say to ourselves at that moment, "Nevertheless, not my will but yours be done."

We may never be challenged with the necessity of sacrificing our lives, but as a leader, sacrifice is inevitable. The mantle of being a leader requires it. No one can force this sacrifice; the willing and courageous choice of the truly transformational leader accepts this fate. This is the final round of the fight and the culminating moment.

These occurrences don't happen often, but when they do, everything rides on the choice we make next. Our credibility, character, integrity, and dependability are all challenged by this choice. When we detach ourselves from the outcome and put the people at the center, we will ultimately win this fight and become the leader our people deserve and need.

The *Fight to Lead* is not just about teaching others but also about leading oneself. It's about the constant battle that leaders must wage against their ego, insecurities, beliefs, self-awareness, livelihood, self-image, fear, and environment. It's about the difficult choices that leaders must make and the sacrifices they must be willing to make to become the leader their team needs.

It's about the journey of self-discovery that leaders must embark on to understand their own strengths and weaknesses and develop the skills and abilities they need to be influential leaders. But it's also about the rewards of leading with integrity, courage, and humility.

It's about the positive impact that leaders can have on their teams, organizations, and the world. It's about the ability to inspire and motivate others to achieve great things and to make a real difference in the world.

The Fight to Lead is not for the faint of heart. It's not for those who are looking for an easy path or a quick fix. You have been through all twelve rounds. You have been brought high, brought low; You have seen the pain and joy of leadership through the examples of others and in the vulnerability of showing what I have learned.

The Fight to Lead is for those willing to work hard and make the sacrifices necessary to become their team's leader. It's for those willing to take on the mantle of leadership and fight courageously.

So if you're ready to take on *The Fight to Lead*, it's time to put on your gloves and step into the ring. The journey will be challenging. You will hit and be hit with round after round, but the rewards will be well worth it.

While in the ring, keep your gloves up and your feet moving. Take to the corner to recover when you are hit hard, but never, never, never, never, never quit.

Remember, at the core, as a leader; you must *overcome yourself* and *become what the people need*—those are your two right hooks of leadership that lead to victory every time.

NOTES

FOREWORD

1. *If you only have a hammer, you tend to see every problem as a nail,* THE FITZROVIA PSYCHOLOGY CLINIC 2018, accessed 1 February 2023, https://thefitzroviaclinic.com/if-the-only-tool-you-have-is-a-hammer-you-tend-to-see-every-problem-as-a-nail/.

INTRODUCTION

1. Âpihtawikosisân, Indian Story, accessed 1 February 2023, https://apihtawikosisan.com/2012/02/check-the-tag-on-that-indian-story/.

CHAPTER 1

1. Ralph Waldo Emmerson. *The Only Person You Are Destined To Become Is the Person You Decide To Be.* Public Domain. 1991 October, Vogue, Volume 181, Issue 10, (Nike Advertisement), Start Page 206, Quote Page 207, Condé Nast, New York. (ProQuest), accessed 1 February 2023, https://quoteinvestigator.com/2020/12/08/destined/.

2. Denis Waitly. Famous Quotes From.com. http://fa-mousquotefrom.com/denis-waitley/.

3. Lazenby, Roland. *Michael Jordan: The Life*. Boston, Massachusetts. Back Bay Books. 2015.

4. Frankel, Viktor. *Man's Search for Meaning*. Boston, Massachusetts. Beacon Press. 1946.

5. Waitley, Denis. *The Psychology of Winning*. Publisher, Melbourne, Australia. Brolga Publisher, 2002.

6. Stockett, Kathryn. *The Help*. Westminster, London, England. Penguin Books. 2009.

7. Hauser, Thomas. *Muhammad Ali: His Life and Times*. London. Robson Books. 2004

CHAPTER 2

1. Zachary M. Schrag. *Abraham Lincoln, First Debate with Stephen A. Douglas at Ottawa, Illinois, August 21, 1858 (excerpt)*, accessed 1 February 2023, https://mason.gmu.edu/~zschrag/hist120spring05/lincoln_ottawa.htm.

2. Susan Segal. *Letter to George Meade July 14, 1863*. House Divided Dickinsen.edu, accessed 1 February 2023, https://housedivided.dickinson.edu/sites/lincoln/letter-to-george-meade-july-14-1863/.

3. Chamberlin, Joshua. *Bayonet Forward—My Civil War Reminiscences*. Publisher: Stan Clark Military Books, Gettysburg. LIMITED, FIRST EDITION; 1994.

4. Segal, Susan. Pinsker, Matthew, *Lincoln Letters. Letter to George Mead*, House Divided, accessed 1 February 2023,

https://housedivided.dickinson.edu/sites/lincoln/letter-to-george-meade-july-14-1863.

CHAPTER 3

1. James Whitcomb Riley. *The Most Essential Factor is Persistence*, Internet Poem.com, https://internetpoem.com/james-whitcomb-riley/quotes/the-most-essential-factor-is-persistence-the-23037/.

2. Julian Harnish, *Can I Tune My Own Piano*, Find Your Melody, June 15, 2020, accessed 2 February 2023,https://findyourmelody.com/can-i-tune-my-own-piano/.

3. Scott Detwiler, *How to Tune a Piano*, accessed 1 February 2023,http://piano.detwiler.us/.

CHAPTER 4

1. Tony Robbins. *Unlimited Power*. New York. Free Press. 2003.

2. Shaara, Michael. *The Killer Angels: a Novel*. New York. Ballantine Books, 1996.

3. *Joshua Lawerence Chamberlain*, Battlefields, accessed 1 February 2023, https://www.battlefields.org/learn/biographies/joshua-lawrence-chamberlain.

CHAPTER 5

1. Kayria, Legson. *I will Try*. Doubleday, Garden City, N.Y., 1965.

2. Barling, Julian, and Julie G. Weatherhead. "Persistent exposure to poverty during childhood limits later leader emergence." *Journal of Applied Psychology* 101, no. 9 (2016): 1305.

3. Bunyan, John. *The Pilgrim's Progress from This World to That Which Is to Come.* 1678. Republish Moody Publishers October 2007.

CHAPTER 6

1. Susan Ratcliffe. *Oxford Essential Quotations (5 ed.). He who learns must suffer.* Aeschylus. Oxford University Press. Online Version: 2017. Agamemnon l. 176. https://www.oxfordreference.com/display/10.1093/acref/9780191843730.001.0001/q-oro-ed5-00000159;jsessionid=C037FAE13CD1935F33B50A0DCBCAB378

2. Ryan Coogler, Black Panther (February 16, 2018; Los Angeles: Marvel Studios), Film.

CHAPTER 7

1. Barrister Chinchilla Person. Heraclitus-2.docx - 9 Heraclitus Heraclitus, https://www.coursehero.com/file/90375198/Heraclitus-2docx/

2. John Grinder & Richard Bandl. *A Book About Language and Therapy The Structure of Magic, Vol. 2.* Science and Behavior Books. 1976.

3. Augustine JR. "Chapter 9: The Reticular Formation," *Human Neuroanatomy.* 2016 Hoboken, New Jersey. John Wiley & Sons (2nd ed.). pp. 141–153.

4. Tony Robbins *Awaken The Giant Within.* 1991. New York, New York. Simon & Schuster.

5. Mandela, Nelson. *Long Walk To Freedom*. October 1, 1995. Boston, Massachusetts. Back Bay Books.

6. William Earnst Henley. *Invictus.*Poetry Foundation.org, accessed 1 February 2023, https://www.poetryfoundation.org/poems/51642/invictus.

CHAPTER 8

1. Tony Robbins *Awaken The Giant Within*. 1991. New York, New York. Simon & Schuster.

2. Earl Nightingale *The Direct Line*. Nightingale.com. Audiobook. Accessed 1 February 2023, https://www.nightingale.com/authors/earl-nightingale/the-direct-line.html.

3. Tom Stevenson. "8 Quotes By Seneca To Help You Lead A Better Life," Medium.com. April 24, 2020. https://medium.com/mind-cafe/8-quotes-by-seneca-to-help-you-lead-a-better-life-f6ddf0cc36da.

4. Nathan Anderson. "The Problem-Solving Problem," Albert Einstien's Quote. Boozell.com. October 12, 2012, accessed 1 February 2023, https://bozell.com/thinking/articles/the-problem-solving-problems/#:~:text=%E2%80%9CWe%20can't%20solve%20problems,it%20comes%20to%20solving%20problems.

5. Everything hangs on one's thinking…A man is as unhappy as he has convinced himself he is" – Seneca.

6. HeartMath Research Center, *SCIENCE OF THE HEART Exploring the Role of the Heart in Human Performance*, Volume 2. Rollin McCraty, Ph.D. Director of Research. https://www.re-

searchgate.net/publication/293944391_Science_of_the_Heart_Volume_2_Exploring_the_Role_of_the_Heart_in_Human_Performance_An_Overview_of_Research_Conducted_by_the_HeartMath_Institute DOI:10.13140/RG.2.1.3873.5128. February 2016.

CHAPTER 9

1. Frankel, Viktor. *Man's Search for Meaning*. Boston Massachusetts. Beacon Press. 1946.

CHAPTER 10

1. John Maxwell. *21 Irrefutable Laws of Leadership*. New York. HarperCollins. 1998.

CHAPTER 11

1. Ralph Waldo Emmerson. *Emmerson's Essays*. Public Domain. Jungle Book. October 2008, https://www.fictiondb.com/author/ralph-waldo-emerson~56935.htm.
2. Dr. Laurence J. Peter. *The Peter Principle: Why Things Always Go Wrong*. Great Britain. William Morrow and Company, Inc. 1969.
3. Robert Holden, "The Happiness Project," accessed 1 Feb 2023, https://www.robertholden.com/the-happiness-project/.

CHAPTER 12

1) Marine Pfc Jacklyn Harold Lucas, Congressional Medal of Honor Website, accessed 2 February 2023, https://www.cmohs.org/.

2) Army Spc John Baca, Congressional Medal of Honor Website, accessed 2 February 2023, https://www.cmohs.org/.

3) Marine Cpl Duane Dewey, Congressional Medal of Honor Website, accessed 2 February 2023, https://www.cmohs.org/

4) Malala Yousafzai. Malala's Stroy, accessed 2 February 2023, https://malala.org/malalas-story

5) Malala Yousafzai – Facts. NobelPrize.org. Nobel Prize Outreach AB 2023. Tue. 14 Feb 2023. https://www.nobelprize.org/prizes/peace/2014/yousafzai/facts/

6) Winston Churchill, Past Prime Ministers, accessed 2 February 2023, https://www.gov.uk/government/history/past-prime-ministers/winston-churchill.

www.ingramcontent.com/pod-product-compliance
Lightning Source LLC
Chambersburg PA
CBHW070546200326
41521CB00024B/659